Making Sustainability Measurable

Michael Wühle

Making Sustainability Measurable

A Practical Book for Sustainable Living and Working

 Springer

Michael Wühle
Hohenlinden, Bayern, Germany

ISBN 978-3-662-66714-9   ISBN 978-3-662-66715-6   (eBook)
https://doi.org/10.1007/978-3-662-66715-6

This Springer imprint is published by the registered company Springer-Verlag GmbH, DE, part of Springer Nature.
The registered company address is: Heidelberger Platz 3, 14197 Berlin, Germany

To all who face sustainable decisions

Preface

The creation of this book has a prehistory that began in 2013 with my becoming self-employed and ended provisionally, but not finally, with the first publication under the title 'Oh je, Herr Carlowitz' in June 2016. The book was the result of my finding phase as a self-employed person in matters of sustainability. In it, I describe the topics, experiences, insights, tools and tricks that I use to be professionally successful.

However, especially at the beginning, it can easily happen that the aspiring sustainability manager wants to cover too many fields of activity and acquires an unmanageable belly shop that is anything but effective. That's exactly what happened to me back then. When I became aware of this, I sorted out and threw overboard a lot of things. This is how the structure of this book came about. I also received valuable tips in this regard from my readers and seminar participants.

The parts that I wanted to continue using professionally had to be sensibly summarized again. The result is what you have in your hand or on your screen right now. Some things from the first edition, such as the topic of cooperatives, which has lost some importance in the meantime, fell victim to the restructuring or were greatly shortened. Other contents such as resource efficiency, energy efficiency, life cycle assessment, carbon footprint and many more found their way into the book—topics that play a bigger role in my professional everyday life as a sustainability manager today than a few years ago.

I expanded the topic complex of sustainability reporting and the EU-wide "non-financial reporting" or also included it for the first time. The previous non-binding nature for companies and organizations is gradually disappearing and I wanted to react to that, as this is a clear sign of the steering hand of politics towards sustainability.

In 2020, I thoroughly revised the book and published it anew in the Springer-Nature-Verlag. Now another two years have passed and I thought that another revision (maybe the last one?) would make sense. Again, I shortened and deleted passages that were no longer relevant, but also added new and very important content such as key figures and values.

However, I was also concerned to preserve the 'spirit' of the book, which had filled me when I wrote the first edition in 2016.

I would like to thank my family and friends at this point, who have always supported me over the years to continue writing about sustainability and to constantly hone and improve my book on it. I am well aware that I will never finish it and that is a good thing.

I wish you a lot of fun reading and trying out my various exercises, as well as the tips and look forward to your feedback.

Hohenlinden   Michael Wühle
July 15, 2022

Contents

About the Author

Michael Wühle, is an experienced engineer and entrepreneur. In his professional career, he has successfully implemented many large technical projects, until he came into contact with the topics of environment, climate and sustainability. Since then, Michael Wühle has made sustainability management his professional focus. As a freelancer, he supports companies and municipalities in the

introduction and implementation of sustainability projects. He likes to share his knowledge in the field of sustainability through seminars, workshops and publications.

1

Definition Sustainability—My Version

What does sustainability mean? The term sustainability has been quite abused and worn out by now and often stands for everything and nothing. We hear statements about our sustainable policy, the sustainable pursuit of peace in the world, a sustainable diet, the great sustainability in the cultural memory, the sustainable development in rural areas, the necessity of a sustainable energy transition, etc. But all this does not explain what sustainability really stands for.

Sustainability is the balanced interaction of different dimensions, which in the classical, business-oriented world stand for themselves alone.
In 1987, the United Nations World Commission on Environment and Development, the so-called Brundtland Commission, published a modern definition of the term sustainability. The name was derived from the chairperson

© The Author(s), under exclusive license to Springer-Verlag GmbH, DE, part of Springer Nature 2023
M. Wühle, *Making Sustainability Measurable*,
https://doi.org/10.1007/978-3-662-66715-6_1

of the commission, the former Norwegian Prime Minister Gro Harlem Brundtland.

"Sustainable development meets the needs of the present without compromising the ability of future generations to meet their own needs."[1]

About 300 years earlier, Hans Carl von Carlowitz had used the term in a new forestry system and thus quasi-predetermined it. More on that later.

This book is aimed at people who face sustainable decisions—be it voluntarily, for example in their commitment to their own community, be it in their professional or private sphere. I would like to bring the meaning, the misinterpretation and the depth of the term sustainability closer to the inclined reader. I am very keen to make clear that sustainability is not an abstract and theoretical term, but rather a powerful tool that can simplify and optimize our actions in practice.

When did I start dealing with the topic of sustainability? I think it was in early 2009, I had just started a new professional challenge as the head of the environmental department of a larger company. One of my tasks was to develop a sustainability strategy for the company and to prepare a sustainability report. It seemed to me at the time as if everything around me was suddenly sustainable. The sustainable development of the company, the sustainability strategy, sustainable fuels, sustainable environmental protection, sustainable reduction of air pollution and noise, sustainable nutrition, everything was sustainable and hip. Sustainability had become fashionable.

Meanwhile, the term sustainability has been quite worn out, as I said, which leads to sometimes tragicomic misinterpretations. If you want, you can also present a battle tank as sustainable. You don't believe me?

Let's go through this for fun: A battle tank works for decades with the greatest precision and destroys in this time all targets (and thus of course also humans) that it is supposed to fight with high efficiency. The economic dimension of sustainability is thus already fulfilled. In addition, these devices belong to the high-end portfolio of the arms industry, they are also produced in Germany, they maintain and create thousands of well-paid jobs. Therefore, one could speak of social sustainability here. What are we talking about anyway? And ecologically speaking? Excellent! Compared to American and Russian tanks, tanks made in Germany certainly have the lowest fuel consumption and the smallest CO_2 emissions during production and operation (e.g. a Leopard 2 battle tank with 1.5 kg CO_2/km^2 is in the same efficiency class as a VW Golf). So a truly sustainable product. Right? It kills people sustainably, that's why the battle tank is sustainable. Correct?

Of course not, but with this somewhat exaggerated example I want to show what capers, twists and perversions the term sustainability has already experienced.

I, however, went about the term sustainability just as carelessly at the time, because the development of my department with all its challenges was more important to me than the interpretation of a word that I thought was only a manifestation of the zeitgeist. Rarely have I been so mistaken in the importance and meaning of a new term for me at the time!

Meanwhile, however, some time has passed and I am firmly convinced that the principle of sustainability is the key to the most pressing problems of humanity. Whether we have in mind the global warming and its consequences for the climate of our earth and all habitats, or whether we think of renewable energies. It does not matter either whether we talk about an intact environment, the preservation of creation, or whether we deal with the necessity of healthy food for all people on this planet. It does not matter either whether we talk about our obligation for the following generations, i.e. about being suitable for grand-children, or whether we denounce the outlawing of child labor and the exploitation of workers in developing and emerging countries—we actually always mean the princi-ple of sustainability. Even if we do not call it that.

The limiting factor in our sustainability consideration is mainly the environment, because sustainability also means to manage the finite resources of our earth. We must not live at the expense of future generations, neither now nor in the future. Therefore, every human being and every organization has the obligation to work on a social development that is ecologically compatible and socially balanced and that serves the economic needs for healthy economic development, which is necessary for secure jobs.

However, if sustainability is actually the key to the most pressing problems of our time, then it has become a little more attractive, rusty and worn-out key by now. It also hardly fits into the matching keyhole anymore. Why this is so, what the key looked like once and how it can become a shiny and easy-to-use tool again, which unlocks the doors behind which the solutions for our problems are waiting, that is what this book is about.

I would like to share my personal experiences around the topic of sustainability and rekindle the discussion about the necessity of a sustainable development. In this way, I hope to make a small contribution to make sustain-ability a living principle again in as many aspects of our everyday life as possible. I will do my best not to argue with a raised index finger or to be instructive.

I would like and have to start with the perhaps most difficult part—the definition of terms. But how do I start writing about the term sustainability without appearing like a schoolteacher? Because, like many complex terms, the term sustainability also offers a wide range of interpretation, depending on location and point of view.

Should I first list what sustainability is not? Should I say, for example, that CO_2 reduction measures are not identical with the term sustainability?

Would it make it easier for us to start if I try to explain why such a misunderstanding of the term sustainability would lead us in a completely wrong direction, or even into a dead end? I could easily write several pages about what sustainability is not, because I also draw from a rich experience. But that does not really help us.

I prefer to try it with the historical roots of the term "sustainability". In doing so, we inevitably come across Hans Carl von Carlowitz, a pioneer of sustainability in Baroque Saxony.

In the Lexicon of Sustainability[3] we read that Carlowitz was responsible for the wood supply of the Saxon mining and metallurgy industry as Saxon chief mining officer around the year 1700. The smelting furnaces of the Ore Mountains devoured huge amounts of wood; in addition, population growth and urban growth led to a great shortage of wood. As in earlier epochs, people did not think further and cut down what was possible. Thus, wood gradually became scarce and thus a great energy crisis arose, which Carlowitz was confronted with and for which he sought and found a solution. He realized that the existing and increasing demand for wood could only be secured by a new type of forestry. With this new method, it could be ensured "… *that there be a continuous, constant and sustainable use / because it is an indispensable thing / without which the country cannot remain in its essence* …".[4] These lines come from his famous work Sylvicultura Oeconomica, which is considered the first independent work on the subject of forestry.

The method of Carlowitz can be simplified as follows: cut down a tree, plant three new trees for it. This was a revolutionary approach for that time, which was not oriented towards the short term, but towards the long term. With this method, with this principle, Carlowitz first of all achieved his primary goal, the assurance of the valuable and "*indispensable thing*" wood.

However, he also achieved two other important things: The regulated forest management that accompanied the implementation of his principle created permanent jobs and thus also a relative prosperity in the affected population. The regulated forest management, in turn, preserved the natural habitats and prevented the karstification and soil erosion, which in turn were prerequisites for the new plantings.

Probably the latter two points were not at the forefront of Carlowitz's considerations. Or did he think of them? Whether intended or unintended, the social and ecological aspects were direct consequences of the new economic key to overcoming the energy crisis.

At the end of the last century, a modern definition[5] was finally created on the basis of the so-called Brundtland Report, which is based on three dimensions:

- **Ecological sustainability:** It is most strongly oriented towards the original idea of not exploiting nature, and thus takes up the thoughts of Carlowitz. An ecologically sustainable way of life is one that uses the natural resources only to the extent that they can regenerate.
- **Economic sustainability:** A society should not live economically beyond its means, as this would inevitably lead to losses for future generations. In general, an economic system is considered sustainable if it can be operated permanently.

- **Social sustainability:** A state or a society should be organized in such a way that social tensions are kept within limits and conflicts do not escalate, but can be resolved peacefully and civilly.

So, now I have actually listed all the essential things that are necessary to determine the term sustainability. Satisfied?

Not really. Me neither. Sustainability is such a complex concept that pure factual knowledge is not enough to understand it. Unfortunately, the term sustainability is not or only partially self-explanatory. My long-time English teacher told me that the English term Sustainability is much more self-explanatory for native English speakers than for German speakers this somewhat wooden and cumbersome word Nachhaltigkeit. I would therefore like to tell you a little story that should enable a simpler and better access to the actual essence of sustainability than a thousand more data and facts can do.

This story is fictional, but it could have happened. Experiences that I have made in recent years in connection with the topic of sustainability are just as present in it as the conditions that Carlowitz probably encountered in the 18th century. I take on the role of a scribe named Felix (who, to my knowledge, did not exist in Carlowitz's life), who assists the venerable Hans Carl von Carlowitz in composing his *Sylvicultura oeconomica* as a student and has the opportunity to ask all kinds of clever and stupid comprehension questions.

We are in the year 1714 in the Saxon Freiberg, in the study of Carlowitz, the central room of a former castle tower, which the Saxon sovereign has given him.

A little history of sustainability

Felix sits in the spacious tower room at a small wooden table and just closes the big book that he has been writing on every day like the last weeks. Today they have finally finished after many months and Felix is glad, because his fingers already hurt from writing so much. He pushes his shoulders back, stretches extensively and looks at the title on the cover again:

SYLVICULTURA OECONOMICA
or
Household news and natural Instruktion
for
Wild tree cultivation

Felix remembers that he had to work for many days for this first page alone and had to start over several times. The letters were more to paint than to write and that was very tedious. But now, at least for today, it's over, end of work.

Something bothers Felix, however. Something is missing for him. He looks up and looks at the evening landscape, which in this early summer is bursting with the green of the trees. It is a green that almost hurts in the eyes. The large windows are partly open and warm air sweeps through the room. The room is filled with drawings and sketches that are nailed to the wall, stand on easels or simply lie disorderly on the floor. They show mine shafts, tools and machines for metal extraction and many, many drawings of trees. Tree seedlings, how they are planted and protected from wildlife browsing. Trees, how they are felled, sawn and stored. Among them also treatises on the appearance and taste of different soils and lists of the number of felled and planted trees. It is a room where obviously a lot of work is done.

In front of the stairwell, the good old tiled stove arches, which on many winter days made writing with cramped and painful fingers just bearable. Next to the stove is the armchair of his master Hans Carl von Carlowitz, uphol-stered with red velvet. As he sits there, holding the pipe in one hand, from which tobacco smoke rises to the ceil-ing, holding a tattered book in the other hand, he looks very relaxed and cheerful to Felix. His master is an old man with 68 years. He has thrown the official wig over the back of the chair. With his round and reddened face, his short gray hair and his noble, but worn-out clothes, he looks more like a university professor than a powerful official of Saxony.

Felix thinks that his master, the highly respected Ober-berghauptmann Hans Carl von Carlowitz, can sometimes be quite irascible and it might be better to say nothing. But he knows now what has been bothering him all the time and not letting his thoughts rest, and so he plucks up his cour-age and addresses his master:

"Master, may I ask you something?"

Carlowitz does not react and Felix repeats his question much louder.

"Can't you leave me alone for a moment to read?", answers the one asked this time.

(Carlowitz had started to mumble more and more with age and so Felix had to strain very hard to listen and often guess what his master had said.)

This time, however, the answer is loud and clear, if not very encouraging. Felix continues anyway: "Master, I have written down your work as dictated by you, but many things are not clear to me and some I do not understand at all."

"That does not surprise me," grumbles Carlowitz, "because I have known for a long time that you are not the smartest. But never mind, today I am generous. What do you want to know, but be brief and speak loud and clear."

Felix nervously pushes the book lying in front of him back and forth on the table top, then opens it and leafs through it aimlessly for a while. Finally he has composed himself enough to ask his first question: "You write in your work that there is a great shortage of wood. That the mines have no more wood for new tunnels, that the smelting furnaces have no more wood to melt the ore into iron. You say that we have to deal with the matter sustainably. What do you mean by that? If there is no more wood here, then we can buy it from the Bavarians or Tyroleans?"

Carlowitz looks at his young scribe with an expression as if he had to explain to a cat what mousetraps are for. He rolls his eyes and looks at the ceiling: "Dear Lord, what a fool have you saddled me, poor sinner, with!", he exclaims. And to Felix: "Haven't you been paying attention all this time? For days, weeks I have dictated to you all the details. Do you just write stupidly or do you also think sometimes?"

With that he slams his fist on the armrest of his chair, making a loud noise. Then he jumps up from his armchair with a quick movement that Felix would not have expected from him. He walks briskly to the table where Felix sits and tries in vain to hide in his chair. In front of the table Carlowitz stops, puts his hands on his hips, glares angrily at Felix and begins to talk to him.

"Once, just once I will try to hammer into your stupid skull what any other person would have understood perfectly well after working with me. So, listen carefully, because if you ask me something so stupid again, you will really make me angry!"

Felix has almost disappeared under the edge of the table and is not able to tell his master that he will listen very attentively. Carlowitz glares at him for a few more moments, probably to check if there are any objections from Felix's side. When he can be sure that he has the full attention of the frightened youngster, he begins to explain his terms in a way that makes Felix suspect that he had prepared this speech for a different audience and had given it several times before:

"As everyone, except Felix, in this country of Saxony knows, there has been a great shortage of wood for many years. Everyone needs wood. We make roofs, tools, carriages, utensils of all kinds from wood, we support the tunnels of our mines with it and above all we need it to

heat our furnaces and smelt our ore. For that is the inex-
haustible wealth of our country. We have, thank God,
enough gold, silver, iron, base metals and minerals in the
rock of our mountains. That is why we have cut more and
more wood to get to these treasures, and in the meantime
most of our forests are bare. Newly planted trees take a
long time to grow and become large—at least as long as
you are an old man—before they can be cut down and
processed. Only with wood for the tunnels and passages
in the mountain, only with wood for melting the ores can
we mine these riches of our country, therefore wood is also
the treasure of our country. We must therefore arrange our
economy in such a way that there is no shortage of wood
and that used areas are immediately rejuvenated. Have you
understood that so far, stupid boy?"

Felix nods vigorously and his master continues.

"Many people now think that the regeneration of the forest can and must be left to the benevolent nature of God
alone. These people doubt the sense of sowing and planting, besides it is more profitable to convert the clearings
into fields and pastures. But the forest seed is nothing
really new, already the ancient Romans sowed and planted
trees in their mighty empire. Without a constant supply
of wood there would have been no Imperium Romanum,
that much is certain. Even now we have supply problems
in Saxony and the wood takes a hundred years to mature.
If then younger trees are cut down out of necessity, this
leads to the devastation and destruction of the maturing
forests.

But as usual in life, people only act when the water is up
to their necks, and that is where we have arrived, because
the wood supply has become very scarce with us. Every
lord, every landowner, farmer and householder should
therefore plant trees everywhere where field cultivation
is not profitable. On the banks of streams and rivers, in
ditches, on pastures and elsewhere. Trees are a treasure and

jewel of a country and the forests are its pantry, but it must be maintained. It takes skill, knowledge and diligence to grow and preserve wood properly, so that there is a permanent, constant and sustainable use, because wood is indispensable and the welfare of the country depends on it.

Imports from other countries like Tyrol, Bavaria or Italy are also not a solution, they would be very expensive, not economical, not sustainable. Moreover, the wood shortage already threatens the whole of Europe. There is therefore only one way—and that is the sowing and planting of trees. If we compare the meagre annual yield from crops in our Ore Mountains with the yield that we can achieve in fifty years from wood, the latter is incomparably higher with many thousand thalers.

Our landowners and businesses have the skill to process wood properly, and our most gracious sovereign will make sure that they are diligent in the matter. The knowledge of how the forest seed works and how trees are sustainably planted, cared for and used, this knowledge we have now written down. It is now available to everyone who wants to cultivate trees and use forests sustainably."

Carlowitz takes a deep breath and looks at his apprentice attentively. "Have you now understood why we are toiling here and what we are working on?"

It is clear that the master now wants an answer from Felix. Unlike before, Felix is not afraid of the answer, because now he understands the connections.

"Yes, master, I got it. Only if we plant enough trees now, then our children will have enough wood for building, heating and mining. And they have to teach it to our grandchildren, so that it always goes on like this. Then we have a perpetual, never-ending source of wealth and prosperity. And we also have a reward in our lifetime from planting and sowing: We can harvest part of the annually sprouting sticks of the young trees, which always grow back, and have our benefit from them."

A gentle smile spreads across Carlowitz's face and he looks satisfied. He knows at this moment that something sustainable had happened here, on this day and at this place. He had seen the seed of his science sprout in his young apprentice and was sure at this moment that this seed would bear fruit.

"Good, good," Carlowitz murmurs, "I think you have understood it now". He walks visibly relaxed back to his armchair, relights his pipe and immerses himself again in the reading of his book.

Felix is also very pleased with everything he has just learned and experienced. But now it is really time to call it a day! Yawning and at peace with himself, he gets up, thinks of the tavern down in the village, the good sausages, the good beer and the pretty innkeeper's daughter, whom he likes to look at. He walks briskly down the tower stairs and thinks no more of his master or of forestry, but of the nice evening with his little pleasures.

That was it, my fictional story about Carlowitz and his tough student. Did you like it? I hope so!

Through my research on Carl von Carlowitz, his time and their challenges, I learned a lot about sustainability. I mean not so much the factual knowledge, but the emotional component, which I try to capture with my little story. This emotional access to sustainability, which our English-speaking friends apparently already have in their mother tongue, this feeling we need to really live and implement sustainability. It is the feeling we have when we put our hand on the bark of an old tree. You know what I mean.

Oh dear, Mr. Carlowitz, one might almost say, when we put ourselves in his time and his problems. He had a gigantic task ahead of him, which required long-term strategies and which had to be planted in the minds of the people.

> It is worthwhile in any case to carry out projects from the perspective of sustainability. Sustainability projects, which ultimately serve all three dimensions, lead to stable, future-proof and thus sustainable results. This applies to products and services as well as to successful transformation of companies into sustainable organizations. This statement is also part of the definition of sustainability.

Well, with that we have now also arrived emotionally in the topic. We recognize and accept that our feelings, our emotions are the indispensable glue that fuse the three dimensions of sustainability into an object, into a unit, which carries a very powerful potential in itself. Carlowitz would not have gotten far with his revolutionary new concept of "wild tree cultivation" either, if he had not won over his fellow men. The command of his sovereign alone would certainly not have been enough. Therefore, we should now turn to the question of why sustainability becomes an indispensable tool in mitigating the consequences in the age of global warming and the associated emotions.

In order for us to answer this question together, we have to keep in mind that people can be most easily taken along on a new path if we can visualize the goal and the result at the end of this path. We thus get a picture of the outcome we are aiming for and can illustrate it vividly.

Source Reference and Notes

1. Our Common Future: Report of the World Commission on Environment and Development, http://www.un-doc-uments.net/our-common-future.pdf, known as the Brundtland Report

2. See http://www.tagesspiegel.de/wirtschaft/leopard-2-so-sauber-wie-ein-golf/4360926.html Whether the information is really true is not so important for our example. What is important is that the argument is similar

3. https://www.nachhaltigkeit.info/artikel/hans_carl_von_carlowitz_1713_1393.htm

4. Sylvicultura oeconomica, Hans Carl von Carlowitz, oekom verlag. http://www.oekom.de

5. https://www.bmuv.de/themen/nachhaltigkeit-digitalisierung/nachhaltigkeit/strategie-und-umsetzung

2

A Picture is Worth a Thousand Words

Do you know the impressive picture that Steve Jobs used to illustrate why Apple has to put equal value on technology and design (art) of the product in order to be unique and successful? Below is a sketch of it (see Fig. 2.1).

Technology and art (or also humanities) meet at a point where a unique product is created. For Steve Jobs, it was extremely important to manifest the compatibility of technology and art in his products.

"Technology, … married with the humanities … that make our heart sing"[1], he once said about it. With that, he nailed the unique selling point of Apple and showed the whole world that not open interfaces to all sides bring the desired stability, reliability and user-friendliness, but closed, coherent and harmoniously cooperating systems produce future-oriented solutions. The prerequisite, however, is that the product is desired by the customer or even better, anticipates the future wishes and expectations of the customers.

Fig. 2.1 Steve Jobs' intersection

And what does that have to do with sustainability? A lot. Very, very much.

Sustainability is also a coherent system, an object, a holistic methodology. If its three dimensions are taken into account in the right way and adapted to the respective situation, then it produces excellent results.

That sounds pretty abstract, doesn't it? I know that and that's why I want to try to illustrate this with some pictures.

Let's look at the picture of the intersection again. It has a weakness despite all its brilliance and the resulting success story, which we also often find in well-intentioned sustainability concepts and which usually leads to failure.

How, weakness? Apple had the mega success with this system, so what's all this talk about weakness?

Yes, weakness.

Anyone who deals with the history of this company, the beginnings and the role of Steve Jobs, will quickly recognize the permanent conflict that almost tore Apple apart several times. It was the conflict between the technicians, who wanted open interfaces and only had the technology in focus, and the designers, who always wanted to create stylish, cool and beautiful products that would be enthusiastically received by the customers. The whole thing was further complicated by the merchants and financial managers, who only had the profit in mind and didn't care about the actual product.

Steve Jobs was the factor that could not only pacify these two, actually three diametrically opposed directions, but also connect them and thus lead to unique successes. Now Steve is unfortunately dead (the gods call those they love early to themselves, or something like that) and I strongly doubt that his successors can and will achieve similar innovations, yes revolutions in the computer and media market. Steve Jobs achieved these successes thanks to his unique genius and his visions.

I don't have this genius, this visionary gift. You probably don't either. Well, you say, we don't need it. We don't deal with computers, tablets, smartphones and the like, but with sustainability, with sustainability management. Our task is completely different.

That is a mistake!

It is also about sustainability to bring together different dimensions, which are all equally important and seem not fit together at all, and to move them to a successful fusion.

Steve Jobs had *only* two dimensions and he managed the happy connection of the two dimensions thanks to his ingenious intuition, vision, power and not to forget: his insane stubbornness

- **Beauty/Aesthetics/User-friendliness/Art**
  (the social dimension of sustainability)

and

- **Technical innovation/Unique selling point/Innovative power/Performance**
  (the economic dimension of sustainability)

For a successful implementation of sustainability criteria in our daily private and professional life, for a future-oriented sustainability management, we have to combine three dimensions—the economy, the ecology and the social/societal dimension.

How on earth are we going to do that?

We achieve this with a clear sustainability methodology that can be applied to every use case, to every project. With that, we ordinary people can achieve great success in the field of sustainability, what Steve achieved thanks to his genius at Apple:

Breaktrough, unique, universally accepted solutions in our daily life, that help us in the present and are also available to future generations.

I will therefore develop a picture with you that is well suited to understand how the different dimensions of sustainability interact and depend on each other.

Before we get to that, however, I would like to illustrate by example why purely social or purely ecological projects often fail if the economic component is not taken into account. These examples, which I have experienced in a similar way, are important so that our picture, our vision of sustainability becomes tangible.

Example 1: Crash of a social project

Let us imagine the following situation: You and I and a lot of other committed people have thought a lot about aging, dementia, care dependency and possible age-related helplessness. This fate can happen to any of us if we live long enough.

All right, we have somehow found each other and want to work together on the solution. Because we want to enable all those affected a dignified aging, living and dying. We also want to provide for our own future and create structures in which we can grow old with dignity.

We therefore plan to found a cooperative that would enable its members exactly this dignified aging. At the founding meeting of our cooperative, we want to decide on the construction of a senior residence, where people can grow old happily and contentedly and be optimally cared for. The senior residence is to be planned and built taking into account all sociological, medical, social and technical findings. It is to become a last home for the elderly, the area is to radiate a serene calmness, embedded in a spacious park.

A great plan.

Now comes the day of the founding meeting. A very important day. The most important day of all, because today the great plan is to get its foundation.

The hall is full. The murmuring points to lively discussions. The atmosphere is expectantly tense. Now the founding mothers and fathers present their great idea, their expectations, their plans to the assembly. Approval from all sides. Great event. Everyone feels happy and secure.

But wait, there is a comment from the audience. Doesn't anyone see that? Yes, yes, now, the stage becomes attentive. Of course, they are happy about every comment. A microphone is passed through the rows and handed to the person in the audience who apparently has something to say.

The person takes the microphone and begins to speak. At first visibly nervous and with a not to be overlooked tremor in his voice, but later becoming more and more confident.

"Great idea you have there, but can you tell me now how you ever want to finance it and what I as a shareholder have from it?"

All of a sudden it is quiet in the audience.

The provisional board on stage takes this very deviating and critical question from the previous course of the event and answers, albeit somewhat hesitantly: "The financing is to be done primarily through equity, i.e. the subscribed shares of the members. If there is still something missing, we will surely get favorable loans."

After that, a several-hour discussion about financial models and return expectations began, which I left at a time when it was still in full swing. My last impression was that it became more and more diffuse, and I really did not have the impression that this evening would bring the desired result.

Believe me, the project was over after this evening before it really started. The reason for this was the lack of preparation and consideration of the economic factors, especially the return expectations of the potential shareholders and other investors.

Example 2: The insolvent energy cooperative

As it happens so often these days, committed citizens found an energy cooperative to supply their members with environmentally friendly and cheap electricity. Let's say this cooperative, which we want to call Morgenrot eG, covers the roofs of their own houses and some other properties of their community with photovoltaic modules and produces solar power.

Full of idealism, the citizens have laid down in the statutes of their Morgenrot eG as a promotional purpose the focus on the production of climate-friendly solar power, to make a contribution to the energy transition. They do not want to make any losses, but also no great profit, a break-even is enough for them. In their business plan, which they have to submit to the cooperative association as Morgenrot eG, they describe the calculated capital return with 1 to 1.2%.

Soon 20 to 30 people from their community are found who think similarly and who join the cooperative. Now there is also enough equity to equip the roofs with PV systems; the concept seems to work. The electricity price is right. The enthusiasm of the members and the voluntary work carries the cooperative. Everything is going great, everyone is happy.

I have experienced this example exactly like this until then.

Now we jump a few years into the future. There are still as many PV systems in operation as before, not a single one more. The capital of the founding members is invested. The very low return on capital did not lead to a capital accumulation, which is not surprising. The money that was left over at the end of the year went into maintenance and repair. New members could not be recruited. All idealistic-minded people in the area had already joined at the beginning. Now the first inverters are due for replacement, because the Morgenrot eG (Morning Red Cooperative) bought mainly cheap products due to its low capital base. With what money should the necessary repowering measures be financed now? In a few more years, the first module changes are due. How is that supposed to be paid for? Should we just switch off the PV systems? That can't be done, that would destroy the invested capital of our members. What to do?

No matter how the story goes on exactly, it will not end well. I bet here on a bankruptcy at the latest ten years after the foundation. The reason for this lies in the clear disregard of the economic dimension. Full of idealism, the people of the Morgenrot eG approached the matter and wanted to make a contribution to reducing greenhouse gas emissions. They worked together great, grew together and still get along very well today. A great community. And bankrupt.

By the way, I recommend for an energy cooperative of this kind a return on equity of at least 3%, better 4 to 5%. Only then can enough reserves be built up to finance necessary renewals and expansions over the years. With such a return expectation, there will probably also be new members in the future who will join the Morgenrot eG. If your business model and your calculations do not yield at least 3%, then I give you the tip: Leave it be!

As a consequence of the two negative examples, I would now like to develop a picture that represents a successful sustainability system. It should be as illustrative as the one by Steve Jobs and cover all three dimensions of sustainability. Because that was the common weakness of the sketched examples. Although in both cases there was a lot of commitment from the initiators and the moral, social and ecological values were fully occupied, these example projects failed quite early, because the economic component was not equally considered. And that in turn came from my conviction that there was no coherent and complete picture of the desired result that all the people involved had in the same way in mind.

This brings us to an interim conclusion: Without an economic approach, without profitability and without a coherent picture of the desired result, ecological and social projects are almost always doomed to fail. If, on the other hand, the economic component is included equally from the beginning, then the chances for a successful project increase enormously.

In contrast to the previously described equality of the three dimensions of sustainability, there is another special feature. In my experience, the economic component must necessarily be the first measure of a project development. Only in this way can the prerequisites and boundary conditions be recognized and taken into account conceptually, so that the actually desired benefit or effect in the social or ecological dimension can occur. Don't you believe me? Well, then try it out and you will return to this point with regret.

But now to our new picture for the desired principle of successful sustainability projects. It is not a crossroads, as with Steve Jobs, but a roundabout. Take a look at this roundabout of sustainability (Fig. 2.2).

From the left comes the economic component and merges with the social component flowing in from below. Both flow together with the ecological component, until they then all unite together take the exit onto the road of sustainability. On the "road of sustainability" they are then united stronger than any single one of them.

I like this picture very much. It shows, on the one hand, that the "road of sustainability" can at least in principle be reached from any dimension. Social sustainability projects often start without the support of the other two dimensions. If, after a few rounds, the other two dimensions have joined, the exit to the "road of sustainability" is still found. Something similar can happen with projects that were initially started on the ecological track.

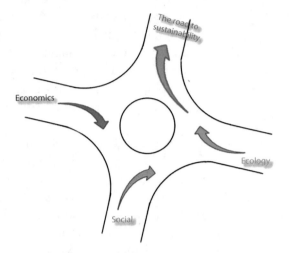

Fig. 2.2 Roundabout of sustainability

Also the economic dimension does not slip through this roundabout of sustainability as easily as a few years ago, because this fixation on the monetary, the financial element is no longer accepted and accepted by the society, by the citizens, the stakeholders and all other interest groups. However, if all dimensions of sustainability unite to form a powerful stream right from the start of a project, then the exit to the "road of sustainability" is found immediately and can be safely driven.

As in any roundabout, all arms are equal and with a common exit, the "road of sustainability", it does not matter in principle who comes from where and who enters the roundabout first.

I would like to relativize the "no role" statement, because I am becoming more and more convinced that, as shown in my picture (not consciously, but subconsciously, my word on it), the social component rightly enters the roundabout as a second dimension, once the economic framework is clear. No matter which sustainability management we will look at later, we always encounter the simple fact that no matter how well-designed economic or ecological concept works in the long run (sustainably), if the human factor is not or not sufficiently taken into account.

It has to do with the enthusiasm and visionary power that lies within each of us. If it succeeds in convincing a group of people of the importance and meaningfulness of a measure, a project, an action in such a way that it appeals to their innermost, their wishes and dreams, then this group can accomplish things that can never be achieved with the usual financial incentive systems and appeals to the responsibility of the individual for the preservation of the environment. Probably each of us has experienced this before and so we can understand how it feels to belong to an unspoken and yet existing spiritual unity.

I cannot explain exactly why this is so, sociologists could probably do that much better. But even if I cannot explain it, you and I can clearly recognize this effect and work with it deliberately in the light of this insight.

Do you remember the story of Felix and his master? Felix's understanding of the system of sustainability and what his master and he had been doing for many months only came to him when Carlowitz told him the core elements so vividly that a coherent picture emerged in Felix's head that was logical and plastic for him. For us, the roundabout of sustainability brings the coherent picture, because it visualizes the equality of the three dimensions of sustainability and the necessity that they unite.

Just as Carl von Carlowitz (and his apprentice Felix) faced the problem of wood shortage and the associated prosperity, we today face the great challenge of a global climate change and its consequences for all areas of society. Carlowitz has already given us the solution approach, we just have to adapt it to our current conditions!

I would like to introduce another picture that we will use again and again in the course of the book. It is a backpack (see Fig. 2.3), in which we will put all the tools and tools that we will work out in the course of the book.

At first this is empty, but you will see, when we come to the end of the book, it will be filled with a lot of useful tools that are necessary for a practical sustainability management and that you can use. The picture of a backpack with the necessary tools I got many years ago in the context of a management seminar and accepted it as useful for me. It is a mental aid, especially in difficult situations.

In such a situation, I recall how many proven and effective tools I have collected and available over the years. Then I choose the tool that is best suited for the respective

Fig. 2.3 Backpack of the sustainability manager

situation and apply it consciously and purposefully. It is as I said only a mental trick, but it works and I can only encourage you to try it out.

Before we start filling our backpack, however, we should first look at the initial situation that leads us to want to apply the powerful system of sustainability in practice.

Source Reference and Notes

1. https://www.goodreads.com/quotes/3191123-it-is-in-apple-s-dna-that-technology-alone-is-not

3

Adaptation to Climate Change and Its Consequences Require Sustainability

It was long denied that we are currently experiencing the beginning of a global warming of our planet with an accompanying global climate change. Meanwhile, these voices have largely fallen silent. Of course, there are exceptions, as we can see from the statements of some Republican politicians in the USA and elsewhere. These are people who are resistant to advice, whom we will encounter more often in this book. They will only accept the existence of an ongoing climate change when their villa in Florida is underwater. For all other people, however, the consequences of the already occurring climate change are unmistakable.

Thanks to the many satellite images, but also through numerous photo comparisons, everyone can now convince themselves in their own environment of how the landscape has changed due to global warming in the last decades. You can find many photos on the topic on the Internet, a good address for this is the website of the

© The Author(s), under exclusive license to Springer-Verlag GmbH, DE, part of Springer Nature 2023
M. Wühle, *Making Sustainability Measurable*,
https://doi.org/10.1007/978-3-662-66715-6_3

Society for Ecological Research e. V. with a very impressive photo comparison of the Zugspitze glacier[1].

I observe new animal species in my home state of Bavaria, which I never saw as a child. Do you know the hummingbird hawk-moth *(Macroglossum stellatarum)*? This is a moth that can be confused with a hummingbird because of its striking flight behavior and that sucks nectar from the flowers with its long proboscis in hovering flight. These moths have become more and more common in our area in recent years, because they can successfully overwinter in increasingly northern regions due to climate warming.

The tree line in the mountains has also shifted significantly upwards. In the middle of the nineteenth century, it was still at just under 2200 m above sea level, but now you can find young growth of stone pines at heights around 2400 m. The so-called subarctic tree line—the area where trees can still grow at all due to the average temperatures—is constantly moving northward.

The permafrost soil thaws with increasing global warming and thereby releases huge amounts of greenhouse gases, especially methane, that were previously bound in the frozen soil. Some researchers now say that this alone will be half of all the carbon dioxide that humans have blown into the atmosphere so far (about 190 billion t) and will further fuel the greenhouse effect.

That the extreme weather events[2] are increasing worldwide, everyone gets to know through the media, who does not consciously look away. Will the famous 2-degree limit be achievable? This limit of global warming, which all reputable climate researchers consider to be the maximum, below which the consequences of climate change remain manageable. And this is a purely fictitious limit, a political compromise. But I think we will soon exceed this limit, too, because the reliable statements of the countries with the largest greenhouse gas emissions on reduction or compensation measures remain rather vague.

But my opinion is not decisive here. Whether the 2-degree limit can be met or not, we all have to deal with the expected and foreseeable consequences of climate change and especially we sustainability managers[3] are in demand here. The mitigation and damage reduction of the unavoidable consequences of climate change, such as extreme weather events, floods and changed flora and fauna must be in our focus. The establishment of a risk management from the perspective of sustainability can be an ideal starting point here.

Therefore, when we advise an organization, a company, a politician or whoever as sustainability managers, we should always ask the following five questions[4] about adapting to climate change:

1. Has the organization established a risk management system?

   Very few companies have a risk management system to cushion and mitigate the risks from climate change. However, it is really not difficult to set one up. Using a SWOT analysis (see Appendix 1) and/or a Sustainability Balanced Scorecard (BSC, see Appendix 7), the risks can be recorded, analyzed and transferred to a risk management system with relatively little effort. The future global and local climate projections should also be taken into account. This is followed by the identification of the risks (but also the opportunities) for the organization. (The SWOT analysis and the BSC now move into our backpack as the first tools.) *From this, suitable measures for damage minimization and risk mitigation can then be derived. The focus here is on the economic dimension of sustainability.*

2. Does the organization consider the impacts of climate change on us and all people?

   This is important for the planning of land and space use as well as for the design/architecture of the infrastructure and buildings and of course also for the maintenance (topics such as the use of greywater, the need for air conditioning, the use of heat pumps instead of conventional heating and cooling systems, etc.). In addition to resource conservation and the use of renewable energies, this addresses the social and societal dimension in particular.

3. Does the organization support regional and global measures to reduce flooding?

   This question will become increasingly important, as heavy rainfall with the corresponding flooding is already increasing enormously and will most likely increase even more. [5] *This also means that natural flood protection such as the expansion of wetlands is of great importance. Along with this, the sealing of surfaces in cities and municipalities must be reduced and we as sustainability managers and consultants of municipalities can do a lot here (see also sect. 6.7, special features of municipalities). The floods of the last years in Germany have shown that so-called century floods are now to be expected every few years. Tamed rivers without sufficient flooding areas will pose an increasing threat.*

4. Does the organization contribute to ecological awareness raising?

   This means raising awareness through education. The aim is to bring people to the realization that mitigating the looming risks from climate change by means of preventive measures is inevitable. It also aims to achieve a corresponding willingness to act by society as a whole. From the awareness that an intact environment is a prerequisite for our well-being, visions and ideas for sustainability projects in all three dimensions then emerge.

5. Are countermeasures initiated?

   Are countermeasures initiated to existing or expected impacts? A contribution is required in one's own sphere of influence, so that the stakeholder groups of the company/organization build up competencies and abilities to adapt.

> Use the checklist in Appendix 8 to clarify these questions and add it to your tool backpack right away. This checklist is part of my sustainability certification for organizations (Fig. 3.1).
>
> For more information on this certification "Sustainability. Now®" please visit my homepage at https://nachhaltig-keit-management.de.

After you have answered these questions for your organization, you can conduct a SWOT analysis. SWOT stands for the acronym Strengths, Weaknesses, Opportunities and Threats. You can also compile the right conclusions and approaches in an extended Sustainability Balanced Scorecard. These are needed to adapt the organization, the company to the specific consequences of climate change. An example of this can be found as already mentioned in Appendix 1.

Fig. 3.1 Example of a sustainability label

Questions lead to answers. How we ask the right questions that inspire our sustainability projects will be discussed later in the chapter on the sustainability manager.

If an organization—be it a business, a municipality, a corporation, an association—has asked itself the questions about the consequences of climate change and has accordingly realigned itself for the future, then this organization has acquired a capability that I like to call fit for grandchildren. I like this term very much, because we can emotionally imagine much more under it than under the term or sustainability. With fit for grandchildren, each of us knows what is meant by that, no further explanations are needed.

The human-caused climate change can, with all its negative impacts, give a strong impulse to companies and states towards sustainability. As a logical consequence, a risk management system is set up at the beginning to make the expected effects of climate change somewhat manageable. Risk management can be an important starting point from the perspective of sustainability to build a complete sustainability management for the affected organization. More on that later.

What applies to organizations also applies in a modified way to each of us. With our consumption behavior and our behavior regarding energy and fuel, we consumers are the decisive force that often only make organizations seriously think about a way towards sustainability.

Source Reference and Notes

1. http://www.gletscherarchiv.de/fotovergleiche/gletscher_liste_deutschland.
2. Detailed information e.g. at the German Weather Service, www.dwd.de.

3. For reasons of better readability, the simultaneous use of feminine and masculine language forms is dispensed with in the following and the generic masculine is used. All personal designations apply equally to both genders.
4. see DIN ISO 26000, Guide to social responsibility.
5. See "Turn down the Heat", http://www.worldbank.org/en/topic/climatechange/publication/turn-down-the-heat.

4

A Change in Consumer Behavior Leads to Sustainability

The establishment of a risk management system as a response to climate change and its consequences is advisable for companies in any case. Whether this is also necessary and sensible for us as individuals must be decided by each of us. For example, if you live in an area that is prone to flooding, you should definitely think about what to do in case of an emergency. But even if a risk management system is not required for us as individuals, our consumption behaviour contributes greatly to stabilising the situation in terms of sustainability.

Never before in our history have we consumers had so much influence on the production and trade of consumer goods. The networked and globalised world enables us to communicate almost immediately, to exchange information about the products we want, we buy and we would like to have. Through our purchasing behaviour and especially through our non-purchasing behaviour, we can literally change the world. Imagine if each of us voluntarily committed ourselves to complying with sustainability criteria in our consumption behaviour. This private sustainability code would shake many entrenched and harmful structures of our society to their foundations.

Let us put together a suitable sustainability code and apply it to our personal consumption behaviour. I use the German sustainability code[1] as a framework, but you can of course also take another one; the essential elements that we need for our personal sustainability code are similar everywhere.

1. PERSONAL SUSTAINABILITY STRATEGY

We consider which aspects of sustainability have a significant influence on our consumption behaviour, which personal sustainability goals we set ourselves and how we can monitor their degree of achievement. In doing so, we take into account in particular the regional value creation that can result from sustainable consumption behaviour.

2. USE OF NATURAL RESOURCES

We make ourselves aware of the extent of natural resources that we use in our household. These include primarily water and energy consumption as well as the amount of waste that we produce annually. We consider how we can use these resources more sparingly and reduce consumption. To do this, we set ourselves qualitative and quantitative goals. If we are able to do so, we calculate our own CO_2 footprint for our family/household and set ourselves goals to reduce the emissions we cause (there are numerous calculation tools for this purpose on the internet, e.g. WWF[2], or at Carbonfootprint[3]).

3. SOCIETY

We consider the impact of our consumption behaviour on the supply chain of the products we consume.

In doing so, we influence the fact that human rights are respected more worldwide and that forced and child labour as well as any forms of exploitation are prevented. We think about what contribution we make to the community in our region or want to make in the future.

> If you now want to tackle the topic of sustainability in your household together with your family, I recommend that you write down these three points for yourself, your family, your household and your regional conditions. Write down the three points and note down below each one what this means for you and your family in concrete terms and what goals you want to set yourself.

Well, now we have set up our own sustainability code. You can probably think of a lot of things that are important and achievable for you and your family. Many of these goals and measures will cost you nothing, others will pay back the necessary investment sooner or later and others you will consciously spend more on if you are convinced of the necessity and can afford it. However, I would like to address some topics at this point and recommend that you take them into account.

- **Energy consumption and energy efficiency**
  You probably already noted this point as one of the first, because it is a lever that you can use immediately and at first also completely without costs. Make a list of the (presumably) biggest energy consumers in your household and think about what savings you could achieve here. With the permanently rising energy prices, we are talking about hard cash that you can simply save in this way. After you have identified the

obviously biggest energy consumers and thought about how you could reduce energy or use energy more efficiently, there are certainly still some consumers that you can hardly estimate. Here, handy energy monitors are useful, which cost little money (devices between 15 and 50 EUR) and which, plugged between consumer and socket, provide you with all the data you need to assess. I use a slightly more expensive device from ELV, but also cheap devices like for example the one from Arendo fulfill their purpose. If these devices are operated for a few weeks on the same consumer, you also get a realistic annual estimate. You will be surprised how much money comes together in your household!

Now record the most important energy consumers in your household and identify avoidable energy guzzlers! To make your work easier, I have included a corresponding checklist in appendix 13, which you can use immediately for your data collection and evaluation of the energy consumers.

- **Renewable energies**
  If you are the proud owner of your own howe, then you may already have your own photovoltaic system. If not, you should seriously consider this in connection with a battery storage. The production costs per kWPeak[4] for PV modules, inverters and meanwhile also for battery storage have fallen significantly below the electricity prices, if you calculate over a 15 to 20 year usage period of the PV system. Maximizing self-consumption and intelligent, house-related smart grids[5], these are the magic words that make the energy transition really become reality. Regardless of state subsidy models, the self-generated energy is more economical than the external supply, if the systems are properly sized and planned. This applies especially to photovoltaic,

photothermal systems, wood chip heating, heat pumps and small wind turbines.

- **Nutrition**

We eat too much meat, if we can afford it. Even though the consumption in Germany is constantly decreasing, everyone still comes to 55 kg of meat per year (https://www.ble.de/SharedDocs/Pressemitteilungen/DE/2022/220330_Versorgungsbilanz-Fleisch.html). This is not only relatively unhealthy, if you want to believe the doctors' incessant warning. It is also questionable from a sustainability perspective. Who has not seen horrible pictures of the excesses of factory farming? Normally we do not look there, because who wants to watch the brutal killings of animals and feel the suffering of the creatures? No, we usually do not want that; we want our cheap meat from the supermarket and otherwise be left alone. I do not want to give a moral sermon, because I am a meat eater and will probably always remain one. However, a reduction of our meat consumption, at least in the long term, could restore the balance here. The high meat consumption also brings other problems with it. The excretions of these huge cattle and pig stocks contribute with the resulting methane emissions (it also stinks) to about 18% of the total emission of greenhouse gases, which are attributable to human activities.[6] The livestock sector is thus responsible for around one fifth of all greenhouse gas emissions, which is about as much as the entire transport sector. And another problem goes along with the enormous meat consumption: the land competition. If more and more land is used for feed production, less land is left for vegetable and fruit farmers. Famines[7] are the consequence and in their consequence impoverishment, flight and wars. According to studies by WWF, over a third of the world's cultivated grain is intended for feeding livestock.

Consequently, a vegan diet would be the way to go, because switching to vegetarian would then be more like a fake package, because for the dairy products that are so popular among vegetarians (and me), the constant "production" of calves is necessary, so that the cows give enough milk. I think the trend is towards a vegan diet, but that does not mean that we all have to become vegans now. If everyone sets personal meat reduction goals in their personal sustainability code and changes their shopping behavior accordingly, that can already make a big difference.

- **Shopping behavior**
One thing that we could all do very easily is to significantly reduce the amount of food that is thrown away every day despite being in good condition, because the best-before date (BBD) has expired. In Germany alone, 18 million tons of food are thrown away every year!

The BBD suggests to us that the corresponding food is bad and inedible after the expiry of the printed date. Even if this was certainly not intended, it happens to all of us that we ignore the milk carton that is at the very front of the refrigerator shelf and take one from further back. We can change this behavior analogously for cheese, butter, yogurt, etc. and thus also influence the production cycles in the long term, which in turn would lead to less "waste".

What I want to say is that we as consumers have it in our hands at this point how much valuable food is produced and thrown away for nothing, even though there is enough hunger and poverty in our affluent western world. In addition to the shelf life of food, the origin and production conditions should also be important for our shopping behavior. Consumer associations regularly check the numerous labels and quality seals that at least give us consumers an indication of whether the product is also acceptable from an ethical point of view. We consumers have it in our hands to change the world for the better in this respect. It would be, if you like, a strategic approach to private consumption behavior.

This attitude towards consumption behavior in the private sector leads to the development of a sustainability strategy for the respective organization in the corporate sector.

Source Reference and Notes

1. See www.deutscher-nachhaltigkeitskodex.de
2. See http://wwf.klimaktiv-co2-rechner.de/de_DE/popup/
3. See http://www.carbonfootprint.com/calculator1.html
4. Kilowatt peak, measure of the maximum power of a PV system or a PV module
5. Term for a *smart* power grid that enables optimization and monitoring of the connected components
6. See https://www.wwf.de/fileadmin/fm-wwf/Publikationen-PDF/Landwirtschaft/Klimaerwaermung-durch-Fleischkonsum.pdf
7. See http://albert-schweitzer-stiftung.de/aktuell/welthunger-entwicklungspolitik-fleischfrage

5

Finding Ways to the Best Sustainability Strategy

We know what sustainability means: Only cut as much wood as can grow back. Live from the yield and not from the substance. With regard to society, this means: Each generation must solve its tasks and not burden them on the following generations. With this statement, which reminds us of Hans Carl von Carlowitz and his fundamental work, the national sustainability strategy of Germany begins.[1]

Yes, only cut as much wood as can grow back. For every felled tree, plant three new ones. I think we agree on this basic principle. For our work, our project in the field of sustainability management, we also need a strategy that is based on the foundations of Carlowitz and adapted and formulated individually for the respective project. The basis of a functioning sustainability strategy is first of all the recognition that the affected company (or society) has to undergo a transformation process that leads away from a purely business-oriented approach to a sustainable form

© The Author(s), under exclusive license to Springer-Verlag GmbH, DE, part of Springer Nature 2023
M. Wühle, *Making Sustainability Measurable*,
https://doi.org/10.1007/978-3-662-66715-6_5

of business. Efficiency and unique selling points are the magic words here.

5.1 Transformation Through Efficiency and Sufficiency

We can all imagine something under efficiency, but what is meant by sufficiency? Keeping the right measure, in my opinion, is a pretty good translation. It also means turning away from the previously compulsively dictated eternal path of growth. The realization that with moderation, with the renunciation of having to consume more and more resources, a satisfied life can also be led, is the basis of any sustainability strategy. In Rio de Janeiro in 1992, the "Rio Process of Sustainable Development" was launched. The international community of states agreed to use the resources of the earth in such a way that all countries of the earth receive fair development opportunities, but the development of future generations is not diminished.

The Agenda 21, the action program for the twenty-first century, was adopted. It contains in 40 chapters recommendations for action for the sustainable management of resources. Ten years after Rio, a balance was drawn in Johannesburg in 2002 and the following topics were brought to the center:

- Resource conservation and resource efficiency
- Globalization and sustainable development
- Poverty and environment
- Strengthening the United Nations in the areas of environment and sustainable development
- Finances
- Technology transfer

Since then, we are still looking for transformation strategies to a sustainable model for states and companies. This also means a departure from the previous, purely business-oriented way of acting. Transformation through efficiency and sufficiency is understood here as a way of life and economy that puts an end to the excessive consumption of goods and energy. This achieves eco-efficiency and consistency, i.e. the compatibility of nature and technology.

5.2 Resource Efficiency Saves Material and Energy

Resource efficiency (RE)—the careful and efficient use of natural resources (in the form of reducing energy, material and water consumption)—is becoming increasingly important for economic, ecological and social processes. Resource efficiency is defined as the ratio of a specific benefit or outcome to the necessary resource input. The Federal Ministry for the Environment[2] describes the goal with the slogan "Making more out of less". This means that growth and prosperity should be decoupled from the use of natural resources as much as possible. This should strengthen competitiveness and reduce resource consumption, thereby reducing the environmental impacts resulting from it.

Resource efficiency has become important for companies of all types and sizes:

- By implementing RE measures, competitive advantages are achieved.
- Resource efficiency is a precursor of a circular economy.

- Almost across all sectors, around every second company perceives the increasing customer pressure and the associated need for more efficiency.
- Due to customer pressure and their own conviction, RE measures are developed and implemented.
- Many companies now see resource efficiency as part of the marketing strategy in their industry.

Resource efficiency can significantly reduce the material and energy consumption in production and processes. This also reduces the production costs and at the same time contributes to environmental and climate protection. Resource efficiency is thus a strategic key competence for all organizations that embark on the transformation path towards sustainability.

5.3   Unique Selling Proposition Sustainability

Sustainability as a unique selling proposition? Yes, because the increased environmental and social awareness of consumers plays a major role for their purchasing decisions. We consumers scrutinize products and services to find out how much sustainability is behind the product/service and its company. Those who deceive their customers with "green" slogans will lose them instead of binding them to themselves.

Companies must know their USP (Unique Selling Proposition) in terms of sustainability and align their strategy clearly according to it—internally and externally. Very often, they forget to communicate this USP to the stakeholders of the company according to the motto "Do good and talk about it".

Sustainability creates value and trust. The concerns about whether sustainable, future-proof action makes lasting business success possible have decreased. Many companies have realized that they can save significant costs by examining their value chains, optimizing processes, reducing waste and focusing on resource efficiency. This behavior towards a circular economy is now closely monitored and positively evaluated by consumers.

Sustainability has become the new business morality. Sustainability is well suited for customer loyalty. It triggers positive emotions and is not a flash in the pan. Those who occupy sustainability issues with their brand, thus create a USP and a competitive advantage.

So far, the prerequisites and preparatory work for creating a successful sustainability strategy.

Building on this, I would like to develop together with you the framework of a sustainability strategy, which you can then adapt and design in practice to your organization, to your project. We work according to the 'Bottleneck Concentrated Strategy' by Wolfgang Mewes. The bottleneck concentrated strategy has four principles and seven phases, which I cannot go into in detail here, but only use the content that we need for our strategy framework. Those who want to deal more closely with the topic, I recommend reading the relevant literature.[3]

As a first step, you should conduct a SWOT analysis of the organization that you want to lead on the way to sustainability. With the SWOT analysis, you find and evaluate the weaknesses and strengths of the organization. Try to bring your SWOT analysis to a single sheet of paper, as shown in Appendix 1. **This forces you to focus on the core issues.** At first, you will probably produce too much text. Cut it down until the content is reduced to the true core (the technique of SWOT analysis is hopefully already in your backpack with the tools and tools !?).

When you have finished your SWOT analysis, then we can move on to the next step. Put the sheet with the analysis in sight on your desk, because we need it more often.

We now want to develop the framework of a bottleneck-focused sustainability strategy. We have already taken the first step with our SWOT analysis. We know the biggest strengths and weaknesses of our organization. We also know what opportunities are available and what risks we have to face. We have already carried out the analysis from the perspective of sustainability. If you have not done that and have done a "normal" SWOT analysis, then go back one step. It is very important that you consider all three dimensions of sustainability.

> **Tip**
>
> Look at the picture of the "Temple of Sustainability" in Appendix 2 and think about which aspects of the three pillars apply to your organization? Which pillar is strong, which is weak? How is the foundation (knowledge of the social responsibility of every organization)? Does the temple have a roof at all (management)?

Now we have recorded the current situation and know in particular our special strengths. Next, we need sustainability goals and a corresponding vision. These will be our guiding principle, which we will use to orient ourselves, with which we can identify and on which we will align all our further actions. The vision answers the questions:

- What are we doing today?
- Why are we doing this?
- What do we want to do in the future?
- What do we want to have achieved in 10, 15, 20 years?

Finding the answers to these questions is certainly not always easy. They must be answers that describe goals at the same time. Goals that inspire us, motivate us and give us the necessary perseverance. **Therefore, take the necessary time.** I have once experienced how a company gave itself a vision on the fly. In the course of a workshop of the executives, the board defined a vision and goals in a coffee break, which were then to be transferred into a sustainability strategy. Setting up this sustainability strategy was still the smaller problem. The implementation failed because there was no discussion beforehand, no possibility to adapt the arbitrarily given goals to one's own ideas. When the board then announced the vision and goals to its executives, the displeasure was very clearly noticeable and there was a lot of protest. Without losing face, the board could not go back and so it remained; you can imagine the result in the implementation.

So take enough time. Discuss an inspiring vision and formulate ambitious and achievable goals!

In our sustainability strategy, the guiding principle could be, for example, the development of a new innovative product, whose entire life cycle from development to recycling process is characterized by implemented sustainability criteria. With such a cradle-to-cradle product[4], a unique selling proposition is also possible and it serves the economic aspect very well.

We could also include the reduction of greenhouse gases in our vision. Reduce so and so many tons of CO_2 equivalents by the year X, which are caused by the activities of our organization. With the introduction of new and energy-efficient technologies, these reduction goals can be achieved. At the same time, we signal to our stakeholders that we are breaking new ground.

Another more strategic approach would be a new structure in the organization of the company, which results in greater efficiency in the processes and a lower energy and resource consumption and allows the employed people a higher degree of flexibility of work and leisure.

There are certainly countless possibilities for an inspiring and motivating vision and you will surely find the right guiding principle for your organization. Maybe you just had a brilliant idea while reading?

At this point, we have to think about what benefits we can derive for ourselves and for our stakeholders from our vision and the goals derived from it. How can we resolve discovered bottlenecks for ourselves and our stakeholders and generate benefits for both sides?

A bottleneck is the thing that prevents us or our customers and stakeholders from developing successfully and positively. If we succeed in resolving our own bottlenecks, then we can develop unique selling points and differentiate ourselves from our competitors. If we know the central bottleneck of our target group, then we can offer them services that have a high benefit for them and that they will then most likely have us provide.

Do you know the picture of the key trunk? In the past, tree trunks were often transported across rivers. The felled and branchless trunks were simply thrown into the nearest river and only fished out many kilometers downstream. Only then were they sawn and processed. It happened again and again that the trunks got stuck together and the transport came to a standstill. Then an expert was called in, who recognized a very specific trunk, the so-called key trunk, based on his experience. When this key trunk was pulled out of the tangled mess, the jam was cleared and all the trunks flowed downstream again. Our strategic approach is similar. If we find the biggest bottleneck at our customer, we can often solve the existing conflicts of an organization with relatively simple measures. It is helpful if you have a picture of your ideal customer (see chap. 2) and identify his biggest bottleneck (key trunk).

> **Tip**
>
> Describe your ideal customer for a sustainability project. It is the one to whom you can offer the greatest benefit based on your strengths and abilities. With the ideal customer, you have also found the most promising target group for your organization.

From the description of your ideal customer, you can now derive a bottleneck analysis, because his obvious problems are now known to you. Then think about what other problems, wishes and needs he might have. Are there growth problems? How is his competitive strength? Does he lack important skills, such as sustainability management? Try to put yourself in your ideal customer as deeply as possible. What worries would you have in his place?

Also think of reasons why your ideal customer has not used the services of your organization so far. Does he know you or your company at all? Is there anything between you? Are there no references? Do you lack advocates? Are your solutions and products well described and easy to find?

Bottlenecks are also always recognizable when emotions are present. Is your ideal customer frustrated because he has a bad image in the public as an energy-intensive or resource-consuming company? Does he have existential fears? Is he clueless and without perspective? Is he envious of the success of competitors? Ask your target group and pay attention to the emotions of your interlocutors. Behind them are the bottlenecks that you can solve for the benefit of both sides.

Now try to formulate the biggest bottleneck that prevents you and/or your ideal customer from doing business together. You should now align your short- and medium-term strategy to this bottleneck. This may also mean that you have to adjust or expand your own vision accordingly. Your goals and the wishes of your target group must match.

> **Tip**
>
> When you have solved the biggest bottleneck, simply repeat the bottleneck analysis and focus on the next biggest bottleneck. This creates a truly sustainable process that protects you from working past the wishes and problems of your target group.

We now know our ideal customer and thus the most promising target group for us. We also know the biggest problem of our ideal customer. Next, we think about what the ideal solution for the biggest bottleneck of our target group could look like and how we could design this ideal solution in accordance with our own sustainability goals and our mission. Keep in mind that you are not looking for a short-term, but for a sustainable solution. Therefore, also look for a possible innovation. Are there new technologies or processes that you could use first? Maybe you have already found this innovation, but it is not accepted by your target group?

Find a partner from your target group who is willing to work with you to apply this innovation towards sustainability for the first time. To win this partner, offer your services at a greatly reduced cost or even for free. Agree with this first customer of your innovation that you can use him as a reference in return. I did this successfully with my certification system for sustainable organizations,

Sustainability. Now.® and learned a lot from the feedback of this first customer. With these insights, I was able to give my innovation the absolute market readiness.

> **Tip**
>
> You can find a partner and first customer for your innovation quite easily at trade fairs. Visit the trade fairs where you expect your target group and your ideal customer. At trade fairs, almost everyone is prepared for news and innovations and therefore open-minded. Just be brave, just talk to people!

As a result of our previous strategic considerations, we can now state the following:

- We know our own strengths and weaknesses (SWOT analysis).
- We have developed a vision and derived sustainability goals from it.
- We have analyzed our target group and found their bottlenecks.
- We have defined our ideal customer and the greatest benefit potential for him.
- We have developed an innovation that can solve the biggest bottleneck.

This can ideally also be the conclusion of our strategy development and implementation. If so, then congratulations at this point! You have strategically and sustainably realigned your organization and are successfully on your way.

Nevertheless, it may be that you do not achieve the desired success alone despite your innovative products and services. You may lack skills or resources that are essential for success and that you do not have. If this is the case, then you should look for one or more cooperation partners according to the motto: "Together we are strong". This or these partners should have complementary characteristics to you and thus create synergies that benefit both cooperation partners.

For example, I had the problem with my organization that we had developed a very innovative product to solve a bottleneck of our target group, but at that time we did not have the financial means to make the product known. We searched and found an organization that had the problem of having disappeared from the public perception due to management errors in the past. Nevertheless, the organization had access to considerable financial resources and still had a large network of potential customers for us. We did not have to talk and negotiate for long—we became cooperation partners and did not regret it. Through our innovative product, our partner became known again in the scene and we were able to market our product successfully.

The key to success here was that both partners were aware from the beginning of the significant and sustainable benefit that could result from this cooperation. Therefore, when looking for cooperation partners, pay attention not only to what your organization can gain from it, but above all and first of all, what benefit your partner will have from it.

> **Tip**
>
> Keep the results of the vision, the goals, your own bottlenecks and those of your target group in writing. Circulate this paper in your organization and review your sustainability strategy regularly, at least once a year.

As the last step, you still have to secure the sustainability strategy of your organization. Nothing is permanent, and the needs, problems and wishes of your target groups will also change over time. You probably know organizations in your environment that initially had a fantastic and innovative product that met the wishes of their target group, but then were overtaken by the respective market development.

The transition from classic mobile phones to the smartphones common today was such a leap that some companies in the industry did not survive. Many underestimated this rapid development simply because they themselves did not have a share in the underlying technical innovation.

For us and our strategy, this means not resting on the successes of the past, but constantly perceiving the changing needs and wishes of our target group. Technology and services change in line with the problems and wishes of our target group. However, their actual basic need, which results from the respective business purpose of the organization, remains constant. We have to align our sustainability strategy to this basic need and adapt it continuously. As long as we can serve the basic need of our target group with our solutions, our business will run successfully.

The consequence of this insight is the need for a think tank in your organization, even if this think tank may be you alone. This position, the person or persons, have the existentially important task of developing the future solutions for the needs of the target groups already in the present. A very good place for this is the sustainability management of an organization.

Source Reference and Notes

1. See   https://www.bundesregierung.de/Webs/Breg/DE/
   Themen/Nachhaltigkeitsstrategie/_node.html
2. See https://www.bmuv.de/themen/wasser-ressourcen-abfall/
   ressourceneffizienz/ressourceneffizienz-worum-geht-es
3. for example: Kerstin Friedrich/Fredmund Malik/Lothar
   Seiwert, The big 1 × 1 of success strategy, Gabal Verlag
4. Cradle-to-Cradle: cyclical resource use, see http://de.wiki-
   pedia.org/wiki/Cradle_to_Cradle

6

The Challenge Sustainability Management

A professional sustainability management is based on the consistent application of all three dimensions of sustainability—the economic, the ecological and the social dimension. The relationships, connections and dependencies of these components are often illustrated with the image of a classic temple.

The Temple of Sustainability

Our 'temple of sustainability' is built on a solid foundation. The foundation of sustainability consists of our expertise through education or studies and the information that is important for our organization, our company, or our private life. On this foundation stand three pillars, one for each of the economic, ecological and social/societal dimension of sustainability. The pillars are equally high, equally strong, i.e. equally important. They support the roof of the temple, the sustainability management. It is directly connected to all three dimensions and thus

© The Author(s), under exclusive license to Springer-Verlag GmbH, DE, part of Springer Nature 2023
M. Wühle, *Making Sustainability Measurable*,
https://doi.org/10.1007/978-3-662-66715-6_6

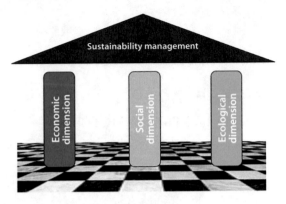

Fig. 6.1 Dimensions of sustainability

ensures an immediate communication among them (see the following fig. 6.1 and appendix 2).

The picture shows very nicely that all elements of the system are interconnected and thus communicate either directly or indirectly with each other. All elements together form a structure that we would immediately call a functional building. The roof of the system, the sustainability management, has another very important function. It forms the common bond, establishes the functionality of the temple and ensures above all the communication between the pillars—i.e. dimensions—of sustainability.

Even though the image of the temple has been used for a long time, it still facilitates the access to the system and the methodology of sustainability. It shows very vividly the necessary equal weighting of the three dimensions of sustainability. But this is only the beginning! The coherent understanding of the term sustainability is the necessary prerequisite for a successful sustainability management. In addition, we have to be aware that we always deal with people. People who (still) do not have the same attitude, the same understanding of the term sustainability.

Sustainability and consequently sustainability management often means drastic changes and many new things, processes and procedures in companies. Many people are afraid of this, it is in our nature. To accept these fears, to deal with them and to reduce them, that is the real key to success. We will come back to this point in more detail later in the book. In the end, however, we have to make the step from theory to practice. This is the meaning of sustainability management.

Well, there it is: our temple of sustainability. It is beautiful, powerful and stable. It is a structure in which all components are related to each other and unite their individual function to a larger overall function. But now we ask ourselves: What use is this ever so stable building, what do we gain from it?

Benefits of Sustainability Management

The unique benefit of sustainability management arises from the structure of three equally important supporting pillars. This ensures that sustainability management will generate positive effects in every dimension. These positive effects also support and strengthen each other.

Let us take a closer look at the benefits that sustainability management generates in all pillars of sustainability. We will start with the economic pillar and this is no coincidence, as I have already shown in Chap. 2 with the sustainability circle.

We still have in mind that probably the social component will be the decisive factor for success or failure, because we now know the power of a mentally connected and united group. Nevertheless, it is wise from a practical point of view to start with the economic benefit. There we speak a language that every board, every CEO, every fiscal officer and every mayor understands. And this is very important to get into the conversation at all. With the

economic benefit—either avoided costs or higher revenues—we can and must arouse the attention and curiosity
of our listeners at the very beginning of every conversation.

It is extremely important to recognize and accept this
priority. This does not mean that I myself start to question
the equality of all three pillars of sustainability. No, not at
all. However, in order to successfully practice sustainability
management, we have to keep in mind the iron principles
of sustainability on the one hand, so as not to deviate from
the path ourselves (and this danger is great). On the other
hand, we have to acknowledge the realities of a purely
business-oriented corporate philosophy in most of today's
companies and align our approach accordingly.

Before I go into the concepts, their measures and their
benefits in the individual pillars, I ask you to take a look
at Fig. 6.2. There you can see a list of overarching benefits
that can arise from sustainability management, regardless
of the specific application area.

Do you know the saying "The fish rots from the head
down"? Applied to organizations, this means: A functioning sustainability management can only be established if
there is a corresponding leadership and corporate culture
in the organization or if it is created.

* Energy and fuel costs are reduced through the use of renewable
energies, alternative fuels and innovative building technology.

* Returns and competitiveness increase through sustainable
procurement (supply chain), innovative technologies and new unique
selling points

* Customers, clients and consumers are more easily won over

* The workforce is motivated and benefits from incentive models.
Attracting and retaining new staff becomes easier.

* Image and attractiveness of the company increases and leads to a
positive perception in the public and in the relationship with all
stakeholders, especially with investors

Fig. 6.2 Benefits of sustainability management in organizations

6.1 Sustainable Leadership and Corporate Culture

The leadership and corporate culture is the shared system of values and norms of a company and shows how a company ticks. It is reflected in how communication is done in the company, how the company is structured, and in the leadership behavior.

The transformation of a company from the classical economic perspective to a sustainable organization is and remains the most important step to be prepared for the future. This has been clearly shown by the Corona pandemic. Sustainability in the company is an effective stabilizer. Sustainable companies are simply less vulnerable to crises and always have a strategic vector directed towards the future. The triad of economic, ecological and social/societal requirements determines the entrepreneurial action. This should be the basis of every modern company. In addition, however, topics such as values, purpose and destiny are becoming increasingly important and should receive appropriate attention.

Values and Destiny
Until now, the principle was that the only purpose of a company should be to generate profits. Here, the sustainability idea has already changed a lot in recent years and especially investors orient themselves more and more to measurable sustainability indicators. The values of a company and its destiny are essential prerequisites for the satisfaction and loyalty of the employees. This also applies to all other stakeholders, especially the customers of a company. They increasingly base their purchasing decision on whether a company credibly lives its values and follows its destiny.

Purpose

To live one's values, to follow one's own destiny and to do this on the basis of sustainability, this is certainly a desirable goal for every human being. It leads to mindfulness and attentiveness. It replaces boring and mindless routine with conscious action that accepts the realities of life. This applies not only to individuals, but especially to every company and its employees. Of course, companies have also already given themselves corporate values (not all) and defined the purpose, the destiny of the company. It is also established that the leaders have to live these values in their daily work in order to bring them to life. In combination with the methodology of sustainability, something new emerges, the purpose. The destiny and the primary purpose of the company. A strong and credible purpose gives the employees and the customers orientation and thus contributes decisively to the positive development of the company. Whether the purpose will replace the previous terms such as CSR (Corporate Social Responsibility), ESG (Environment, Social and Corporate Governance) and other terms used in the context of sustainability for companies, I cannot say. But it is a new approach that expands the previous access to sustainability in companies and definitely has the potential to act as a corporate guiding principle in the next years.

For a company that wants to install and integrate a sustainable leadership and corporate culture, the fields of action corporate mission statement and symbolic management are important:

- **Corporate mission statement**
  Summary of the action-accompanying values and orientation points that provide the employees with a framework for thinking and reference. Sustainability as part of the mission statement fulfills an important

orientation function in this context—especially in ecologically and socially relevant decision situations. However, the best mission statement remains insignificant if it is not lived by the leadership on a daily basis.

- **Symbolic management**
  The conscious design of the so-called corporate artifacts. These include narratives that circulate in the company, typical ways of speaking, established rituals, etc. It reflects the attitude towards the environment and the employees.

- **Appreciative communication**
  Publication of sustainability success stories in the company. Regular joint actions and competitions of leadership and employees on the topic of environmental and social responsibility. This way, actions follow the words and postulates of the corporate management.

Now we can and should go about formulating strategic guidelines for our organization and deriving sustainability goals from them, based on our leadership and corporate culture.

6.2 Strategic Guidelines and Objectives

Every consciously developed organization builds itself on the basis of a vision and develops strategic guidelines to formulate this vision as clearly as possible. A vision is always also linked to objectives, even if not all objectives are quantifiable at the beginning.

For example, a freight company could develop the following vision for itself: "… We want to be the most sustainable forwarder in Europe by the year 20xx, reduce our

CO_2 emissions by 30% and convert all our trucks to electric drive …"

Strategic guidelines and objectives have a tremendous impact internally and externally, for better or for worse. Therefore, it is important to define ambitious, but also achievable goals. Then the workforce and all other stakeholders identify with the vision and the goal.

With this, we are now conceptually well prepared. Our sustainable leadership and corporate culture knows, in addition to the requirements for leadership behavior, the decisive power of the social dimension, the necessity for ecological behavior and the needs of an economy oriented towards consumption and profit. From this we have derived our strategic guidelines and defined our sustainability goals.

With this knowledge, we can move on to defining measures for our organization. To do this, we sort the dimensions of sustainability into a non-judgmental order that promises the highest possible probability of realization. We therefore start building our sustainability management by formulating the economic concept.

6.3 Concept for Economic Behavior

First, we arouse the interest and attention of our interlocutors by explaining the economic potentials of a sustainability management. We do this so vividly that our conversation partner can literally hear the money clinking that they can save or generate additionally. Let us look at the possible measures and the economic benefits of an established sustainability management in a company or in a municipality in Fig. 6.3:

This table is only an example of many possible ones for a measure/benefit table. For your organization, you have

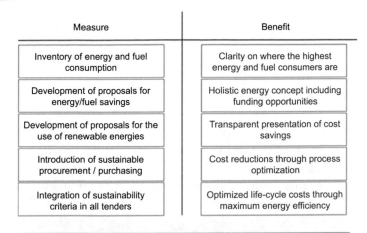

Measure	Benefit
Inventory of energy and fuel consumption	Clarity on where the highest energy and fuel consumers are
Development of proposals for energy/fuel savings	Holistic energy concept including funding opportunities
Development of proposals for the use of renewable energies	Transparent presentation of cost savings
Introduction of sustainable procurement / purchasing	Cost reductions through process optimization
Integration of sustainability criteria in all tenders	Optimized life-cycle costs through maximum energy efficiency

The use of renewable energies offers the opportunity to save considerable energy and fuel costs or to generate revenues. This is supported by optimized processes in the administration

Fig. 6.3 Economic benefits

to select what is applicable for you and add what is missing. However, you can follow along this example, because most of the points will also apply to your organization. Let us now take a closer look at this example. The first line may seem trivial to us at first: inventory of energy consumption. Sounds boring and useless, but it is not. In my experience, few entrepreneurs and municipal officials know how much energy is consumed in their area of responsibility. Even fewer know what the energy is used[1] for and what costs it causes in relation to the total costs. This also applies to the private sector. Could you answer these questions about energy consumption for your apartment, for your house? Most likely not!

What hurts the wallet in the private sector, that is usually much more painful for a municipality or a company. Because let's be honest. In the private sector, everyone looks more or less at saving energy. In the company, at the workplace, this is still much less important for many people.

So, we now know about our energy consumption (you possibly for the first time in your private sector). The benefit we can draw from this is a completely transparent overview of our energy costs, broken down to the last kilowatt hour. I advise to be as fine-grained as possible at this stage. Break down your listing by electricity, heat, cold and fuel consumption.

Assign these consumptions to your organizational units, your properties, your locations, or other groupings that make sense for you. I like to work with a spreadsheet program at this point. It is simple, fast and clear. With the various filter functions, I can then get a picture from different perspectives after the inventory. First, you should sort by the amount of energy consumption and then move on to the next step.

This consists of finding potentials for significant savings and developing concepts from them that already have a clear practical relevance. It is very important to make sure that these concepts are not theoretical, general and academic. If statements like "… we all have to save energy …", "… we should print less …", "… we should really consider whether we need this business trip or not …" come up at this stage, then it is high time to stop them.

We have to develop concrete concepts in this phase and calculate their saving volume. It makes sense to assign the potentials we have found to the order structure that we applied in the inventory. At the same time, we should also differentiate our measures according to:

- Measures for energy saving and
- Measures for energy efficiency.

So, for example, if we have structured according to real estate, then we assign the measure "passive cooling" in the category "energy saving" to the buildings where this measure is feasible according to our first assessment.

In this phase, we should also inform ourselves whether and if so, there are funding opportunities for this at the state, federal and EU level.

After this phase, we have developed a holistic energy concept for our municipality or our company and can confidently present the results to the board or the mayor.

If we have done our work well and the board or the mayor are not completely resistant to advice (which can happen!), then packages of measures will be put together and implemented. What to consider in this process, I will go into more detail in the chapter on the sustainability manager.

First of all, we have incorporated what we can save or use more efficiently in energy into the economic concept. Now we look at how renewable energies can be used in the company or in the municipality. At this point, our focus is still on the economic aspect. Which renewable energies can we use that cost less than fossil fuels?

There are none, you say?

There are, we just have to find them. I claim at this point that I can find at least one energy source from renewable sources for every company location, for every municipality, that causes lower specific costs under life-cycle consideration, and you can do that too!

A considerable economic potential also offers the purchasing. Here, sustainable products can be procured at at least the same, often lower costs. Almost all managers I have talked to about sustainable purchasing so far have immediately objected that these goods would always be more expensive than standard products. But that is not true!

Look at companies like MEMO, Edding and many others. You will find that products with a very good environmental balance and fair production conditions are gaining more and more attention and buyers on the market. We consumers already exert quite a lot of pressure, tendency rising. A company, a municipality, that orders in much higher quantities than a private consumer, of course has an even greater influence on the market for sustainable products and should use this consistently. It is not that difficult either.

First of all, to the term: Sustainable procurement—what is that?

Sustainable procurement is the process of procuring products that have lower consequences for people and the environment than comparable products and services from production to disposal, taking into account social, ecological and economic aspects. Ideally, a circular economy is set in motion through sustainable procurement. In the context of procuring services, we understand the observance of our ethical basic ideas in the sense of sustainability.

The procurement process is subject to certain guidelines in every organization, which are more or less extensive depending on the size and enthusiasm for administration and bureaucracy. In any case, however, various criteria are applied to a procurement. Of course, the price is relevant and things like availability, quality, delivery time. Sometimes, an environmental or an energy certificate is also valued. The existing purchasing criteria only need to be supplemented by a new list of sustainability criteria, which are weighted and taken into account in the decision-making process. An example of what such a list can look like is attached in Appendix 3.

The sustainability criteria that we now consider in purchasing can then also be transferred to all tenders for investment measures. These are things like environmental and sustainability certificates, which are demanded and rated, proof of the renunciation of child labor for internationally active companies, short transport routes and preference for local products, low energy consumption and low-energy production, etc. For new building construction, for technical systems and plants, optimized live-cycle costs can be achieved, which make every businessman's eyes shine.

Conclusion: Especially the use of renewable energies offers the chance to save considerable energy and fuel costs or to generate revenues. This can be supported by optimized processes in the administration. The emphasis on this positive financial effect must be put in the foreground at the beginning, in order to enable a comprehensive sustainability management in the affected organization at all.

6.4   Concept for Social Behavior

In order to create a holistic, a sustainable concept from the economic behavior, we have to examine the social/societal behavior closely and integrate it into the overall concept. Because as we have already discussed at the beginning of the book, the social, the societal element is a central key point and indispensable in harmony with the other dimensions of sustainability.

Only with motivated and convinced people can we build a sustainability management that is stable, future-oriented and sustainable in the true sense of the word. That is why ideally the social/societal component follows the economic component. We have already established this at the beginning of our considerations on sustainability with the sustainability circle.

Here too, I would like to start with an example of what such a concept could look like for any organization. It is only one example of many and you should of course create something of your own and absolutely fitting for your organization. The important thing is that you always make sure that the respective measures also have a real benefit for the target group. Without creating any benefit, you will not have any success. Later I will go into more detail on this indispensable necessity of generating benefit. Let us now take a closer look at the measure/benefit table of the social/societal dimension of sustainability in Fig. 6.4:

Our table starts with staff and citizen models for the use of renewable energies. If we build a photovoltaic system on the roof of a factory hall or a town hall to reduce the energy costs, then we can and should let our staff or citizens participate in it. Only through this benefit for each participant does the commitment arise, which is necessary to keep the process permanently (sustainably) alive, over a often already existing positive basic attitude towards sustainability.

Here too, you should ask the following (and more of your own) questions and translate the answers into further actions:

- Employment and employment relations
    - Does the organization ensure that all work is performed by women and men who are either legally recognized as employees or self-employed?
    - Does the organization conduct active workforce planning?
    (To avoid casual or excessive temporary work.)
    - Does the organization inform in a timely and appropriate manner about planned changes in the operational process, together with representatives of the workers?

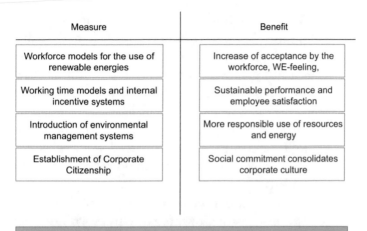

Measure	Benefit
Workforce models for the use of renewable energies	Increase of acceptance by the workforce, WE-feeling,
Working time models and internal incentive systems	Sustainable performance and employee satisfaction
Introduction of environmental management systems	More responsible use of resources and energy
Establishment of Corporate Citizenship	Social commitment consolidates corporate culture

Involving the workforce in the company's sustainability concept promotes a sense of "WE" and increases identification with the company's goals.

Fig. 6.4 Social benefit

- Does the organization ensure equal opportunities for all workers and prevent direct or indirect discrimination?
- Does the organization prevent any kind of arbitrary or discriminatory dismissal and termination practices?
- Does the organization protect the personal data and privacy of the workers?
- Does the organization ensure that work is only assigned to contractors or subcontractors who are legally recognized and provide decent working conditions?
- Working conditions and social protection
  - Does the organization ensure that the working conditions comply with the national laws and regulations and are consistent with the international labor standards?

- Does the organization ensure that the minimum provisions for labor standards of the ILO are respected?
- Does the organization offer equal pay for equal work?
- Does the organization offer working conditions that allow a balanced measure between work and private life?
- Does the organization respect the right of workers to adhere to the usual or agreed working hours, as established by law, regulations or collective agreements?
- Does the organization respect the right of workers to form or join their own organizations with the aim of promoting their interests or conducting collective bargaining?
- Health and safety at work
  - Does the organization develop guidelines for occupational health and safety and ensure their implementation and compliance?
  - Does the organization analyze and control the health and safety risks associated with its activities?
  - Does the organization provide the necessary safety equipment, including personal protective equipment?
  - Does the organization document and investigate all health or safety related incidents, in order to reduce or eliminate them?
  - Does the organization offer appropriate training for the entire staff on all relevant topics?
  - Does the organization involve the employees in the health, safety and environmental program?

The answer to these questions will certainly give you a good basis to assess and optimize the social behavior in your organization towards sustainability. However, do not forget that this is about people. Talk to the workforce, talk to all stakeholders, so that you get the right gut feeling for what is really necessary and where the shoe really pinches, besides all the facts. Give people the opportunity to contact you confidentially and anonymously, because as a sustainability manager, this basic mood is the foundation on which you have to build and implement your sustainability concept.

In the social dimension, we also have to be aware that the so-called "social innovations" are the elements that make outstanding and revolutionary technical innovations possible. This is one of the most important insights in the field of corporate social responsibility (CSR) or sustainability. First, the social behavior, the social demands have to change! Only then follows the technical innovation.

If we look at the development away from desktop PCs and laptops to smartphones and tablets, then what happened there is for me a prime example and proof of my thesis. We simply did not want many different devices to communicate, listen to music, watch videos, access data. No, we wanted a handy, appealing and intuitively usable device, and with this social demand we provided the impulses that made brilliant people like Steve Jobs invent something great and useful like an iPad or an iPhone. First comes the change in social behavior and social demand, then follows the technical innovation and not the other way around.

6.5   Concept for Ecological Behavior

We have overcome the first two hurdles. With hard economic facts, we have conjured up a first glimmer of enthusiasm on the cheeks of our executives and mayors, and we have also emotionally reached the employees and stakeholders. Now the dimension of ecological sustainability has to be equally integrated into the ongoing process. We know that this ecological pillar is wrongly often confused with the whole sustainability and therefore we consider it only after the economic and social dimension, in order not to let this misinterpretation arise in our minds at all.

We do not want to forget and remind each other again and again that the temple of sustainability can only stand and function securely if its three pillars are equally large and equally strong. Because only under this condition the pillars support each other optimally. Remember, we put the economic pillar on the manager's desk first for tactical reasons and had success with it. They listen to us and start to follow our reasoning and advice.

Next, we built up the social/societal pillar and that also happens for the well-considered reason of taking the affected people along. As soon as the economic key figures are clarified and the executive board or mayor has realized that we are talking about considerable cost savings through process optimization and more efficient energy use, even about profits from the generation of energy, we now have the attention to highlight the benefits of a changed ecological behavior.

I have often experienced that executives and mayors at this point of the discussion enthusiastically report on ecological projects they have heard of, and how good they are for the environment, etc. I am always surprised by the factual knowledge that these managers often have, which they have acquired privately or in seminars. Then I feel the desire to participate in ecological projects, which they could never do before, because environmental projects, green projects (yuck), are always deficitary! And then someone comes along and shows them that this is not the case and that these projects are also good for the people and the working climate. Well, if that's the case, the managers think, then of course we can talk about it!

There are numerous definitions for ecological sustainability. I personally like this one best:

The term ecological sustainability describes the far-sighted and considerate use of natural resources.[2]

It is about the survival and health of ecosystems. A neglect of ecological sustainability leads to the irreversible destruction or uselessness of certain resources and thus destroys the chances for any further developments.

Let us now talk about the concept of ecological behaviour within the framework of a functioning sustainability management. Let us look at the following table of typical measures from the ecological dimension in Fig. 6.5 and their benefits for a municipality or a company. By the way, the contents of this table also do not claim to be complete, quite the contrary. They are intended to show by way of example what kind of beneficial measures can be achieved in the three dimensions of sustainability management. For your specific project, for your company or your municipality, you have to adapt and extend these catalogues of measures to the specific conditions. However, the examples show you important key elements.

So, here is the measure/benefit matrix of the ecological pillar:

A classic step at this point is the creation of a CO_2 inventory, the famous CO_2-footprint, which the emissions of greenhouse gases caused by us leave on the beach of the climate ocean. This footprint is decisive for our further approach. We can develop strategies, derive measures and carry out projects based on this footprint. Due to the importance of a complete and correctly calculated CO_2 footprint, I will go into this aspect of the ecological pillar of sustainability in more detail.

First of all, the general term CO_2 footprint is a bit misleading. What is actually meant here is the footprint of all six defined greenhouse gases from the Kyoto Protocol,

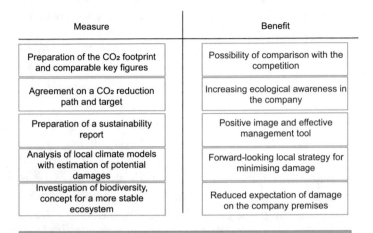

Measure	Benefit
Preparation of the CO_2 footprint and comparable key figures	Possibility of comparison with the competition
Agreement on a CO_2 reduction path and target	Increasing ecological awareness in the company
Preparation of a sustainability report	Positive image and effective management tool
Analysis of local climate models with estimation of potential damages	Forward-looking local strategy for minimising damage
Investigation of biodiversity, concept for a more stable ecosystem	Reduced expectation of damage on the company premises

The agreement of greenhouse gas reduction targets and the analysis of local ecological conditions increase environmental awareness and the readiness for concrete environmental projects.

Fig. 6.5 Ecological benefit

converted into so-called CO_2 equivalents. According to current knowledge, there are six greenhouse gases that were first summarised in this context in the Kyoto Protocol. They are the gases CO_2 (carbon dioxide), CH_4 (methane), N_2O (nitrous oxide), H-FKW/HFC (partially halogenated fluorocarbons), FKW/PFC (perfluorinated hydrocarbons) and SF_6 (sulphur hexafluoride).

These greenhouse gases are converted into CO_2 equivalents taking into account their different climate effects and summed up. This gives us a reference value that is easy to calculate with. For example, the CO_2 equivalent of methane is 25 over a period of 100 years. This means that one tonne of methane has the same climate effect as 25 t of CO_2. If you want to know more about the topic, you can quickly read up on the most important facts on the Internet, for example at https://klimaohnegrenzen.de/.

The example concept shown in the previous figure focuses on the problem of climate change and the greenhouse gas emissions that cause it. This is a very popular and also successful approach, as it addresses an environmental issue that many people deal with and that is well suited to create a corresponding awareness. In a municipality as well as in a company, an authority or any other organisation. With the topic of climate change and its consequences, we thus get a relatively easy entry point for ecological behaviour. However, we must also be aware that there are a lot of other ecological issues that we have to look at just as intensively as the emission of greenhouse gases, depending on the specific circumstances.

We should therefore ask the following questions to the organisation concerned, whose answers can form the basis for a future sustainable and comprehensive ecological behaviour:

- Air pollutants and water pollution
  - Does the organisation emit air pollutants?
    If yes, which ones and how much of them?
  - Does the organisation cause discharges into water bodies?
    (This refers to water pollution caused by direct, intentional or accidental discharge into surface water or by seepage into groundwater.)
  - What is the specific wastewater volume of the organisation?
  - Has the organisation initiated measures to reduce air pollutants and to protect water quality?
- Waste management

- What is the specific waste generation? Are there any hazardous wastes and what is the proportion of hazardous wastes in the total waste generation?
- What is the recycling rate, i.e. the proportion of waste for recycling in the total waste generation?
- Does the organisation sell waste to other countries? (This means whether waste is sold to other countries for disposal or for recycling.) If yes, are there any developing or emerging countries among them?
- Are local communities involved? (This means the involvement with regard to actual and potential pollutant emissions and waste, to corresponding health risks and to current and planned measures to mitigate them.)
- What measures has the organisation taken to reduce waste?
- Minimisation of risks for human and environment
  - Is there a hazardous substance generation in the organisation? If yes, what is the proportion of hazardous substances according to the country-specific hazardous substance regulation or similar regulations?
  - How many (reportable) environmentally relevant accidents, incidents or damage events occurred in the past years?
  - Are there any environmentally relevant hazards and risks?
  - Are there any potential or special hazards and risks for human and environment due to technology, production, transport or other unknown impacts?
  - Does the organisation pursue a responsible approach to technologies?

(This means the use of nature-based, fault-tolerant, reversible technologies with low intervention depth, low and predictable risk potential.)
- Is there a programme to prevent and manage environmental accidents?
- Material use and water consumption
  - What is the consumption of raw, auxiliary and operating materials?
  - What is the material share for product and outer packaging of the products that the organisation produces or uses itself?
  - What is the specific water consumption of the organisation?
  - How large is the freight traffic volume?
  - How many tonne-kilometres on road, rail, plane or ship transport are caused by the organisation?
  - How many business trips are there in the organisation? How many business travel kilometres on road, rail, plane or ship transport result from this?
- Mitigation of climate change → Greenhouse gases
  - Is there already a definition of the system boundaries for the organization?
    (e.g. boundaries of the factory premises, including all routes driven by transporters)
  - What is the total emission of greenhouse gases (GHG) in CO_2 equivalents?
  - What are the direct and indirect GHG emissions?[3]
  - Are there measures to reduce the direct and indirect GHG emissions in the organization?
- Mitigation of climate change → Energy saving
  - What is the total energy consumption of the organization for electricity, heat and cold?
  - What is the fuel consumption of the own fleet?
  - What is the share of renewable energy sources?

- What measures for energy saving and increasing energy efficiency has the organization introduced?
- Adaptation to climate change
  - Has the organization established a risk management? (to take into account the future global and local climate projections and identify the risks for the organization)
  - Does the organization consider the impacts of climate change? (in the planning of land use, land cover and design of infrastructure as well as maintenance)
  - Does the organization support regional measures to reduce flooding? (This includes the expansion of wetlands for flood protection and reduction of soil sealing in urban areas.)
  - Does the organization contribute to ecological awareness raising? (Raising awareness through education to recognize the importance of adaptation and preventive measures. Bringing about a corresponding willingness to act in society.)
  - Are countermeasures initiated? (Initiation of countermeasures to existing or expected impacts. Contribution in one's own sphere of influence, so that stakeholders build up competencies and abilities to adapt.)
- Environmental protection and biodiversity
  - Does the organization identify negative impacts on the environment caused by its activity? (This means that potential adverse effects on biodiversity and ecosystem performance should be identified and measures taken to eliminate or minimize these effects.)

- Does the organization participate in the costs of climate change?
  (This means the participation of the organization in market mechanisms to internalize the costs of its environmental impacts and create economic value by protecting ecosystem services.)
- Does the organization give the highest priority to the conservation of natural ecosystems?
  (Highest priority to the conservation of natural ecosystems, followed by the restoration of ecosystems. Creation of additional green spaces related to nature conservation that go beyond what is legally required)
- Does the organization use sustainable products?
  (the organization should gradually use more products from suppliers that apply more sustainable technologies and processes)
- Does the organization protect natural habitats?
  (this includes the development of construction measures as well as building biological, aesthetic and humane principles)

This questionnaire should be completed in any case in order to create a concept for ecological behavior of the affected organization. You will find that by answering these questions alone, a process will emerge within the organization that can serve as a driving force for the entire sustainability management. I have often experienced at this point that it suddenly clicks in the minds of those people who deal with these questions, and that these questions trigger a thought process that deals intensively with the enormous potentials from the ecological pillar.

The questions listed here and many more have been integrated into a self-developed questionnaire with which I can check any organization for compliance with sustainability criteria and certify it with my own quality seal

Sustainability. Now.®. Recommendations from the ISO 26000[4] were also taken into account. If you want to learn more about this quality seal and maybe use it, you can find out more about it at http://www.nachhaltigkeit-management.de/.

The concept for ecological behavior takes up a lot of space within sustainability management, as so many questions have to be asked and answered in order to derive the right actions and measures. You probably noticed that there are a lot of overlaps between the economic and ecological dimension of sustainability, especially in the energy sector, but they support each other. This can be used well in the argumentation in the sense of "reducing energy costs and doing something good for the environment at the same time".

After we have now made considerations for corresponding concepts for all three dimensions of sustainability, asked many questions and received answers, we are almost ready to look at the different organizational forms regarding special features in terms of sustainability. Before we do that, however, we should devote ourselves to a topic that, no matter what organization we are talking about, is becoming increasingly important in the context of sustainability—sustainability in the supply chain.

6.6   Supply Chain Management— Sustainability in the Supply Chain

The chapter on sustainability in the supply chain would actually fill a book of its own. The topic is very essential for the practice of every sustainability manager and therefore cannot be missing here. I will touch on the key points at least and refer to important sources. However, in order for the interested reader to be able to delve deep enough into the topic, I will explain the sustainability in the supply chain using a small practical example, to which I invite you to play along. In my seminars on the topic of supply chain, I have had very good experiences with this exercise. It is about the procurement of tires for a company with a large fleet of vehicles. Just let yourself in, it will surely be fun and trigger some aha effects!

But first some theory: Supply chain management—what is it?

The roots of supply chain management (SCM) grew in the USA in the early 1980s and the topic, like so many others, spilled over to Germany in the mid-1990s and has been gaining importance here ever since. However, SCM is also becoming more and more "green" and is often misunderstood. Caution is advised here, because in SCM, too, all three dimensions of sustainability must always be considered equally, otherwise the invested efforts will fizzle out here as well.

Sustainability in the supply chain begins with the promotion of good corporate governance and leads through purchasing and production to use and subsequent recycling (see also the following Fig. 6.6). Sustainability in the supply chain is part of a circular economy over the entire life cycle of a product.

The potential for financial savings is often the bridge to SCM and is therefore put in the foreground at the beginning of an internal discussion. We must not forget that SCM represents the sustainable management of the environmental, social and economic impacts over the entire product life cycle and the possibility of reducing costs is an unbeatable argument for its introduction.

For goods and services, a sustainable supply chain looks as follows:

Fig. 6.6 Aspects of a sustainable supply chain

- Research and development (innovations)
- Purchasing/procurement
- Production (goods), infrastructure (services)
- Product use (goods), service
- Logistics
- Recycling (goods)

The infrastructure of a service is, for example, the hospital for the nursing service or the kindergarten for the educator. The development of innovative products and services can be a unique selling point for the organization and thus provide a competitive edge over competitors.

At this point, I would like to refer to the Global Compact Network and recommend you to read the background paper Innovation and Sustainability[5]. The United Nations Global Compact is a strategic initiative for companies that commit to aligning their business activities and strategies with ten universally recognized principles in the areas of human rights, labor standards, environmental protection and anti-corruption.

Well, that's enough theory for now. Time for the first exercise, which you can do alone or with several participants. For a larger number of participants, you should form groups of three to four people each.

Exercise 1
Are you ready? Then go back a few lines and memorize the individual elements of a sustainable supply chain. Imagine that you have been commissioned by the board of your organization to procure a new type of car tire for your company's fleet. Sketch a possible supply chain for the procurement in about ten minutes and consider the following boundary conditions:

- Assume that the suppliers are located abroad.
- Consider the complete process chain from research and development to recycling.
- Write down the main points of the supply chain on cards or notes.
- Now arrange your supply chain in a meaningful order and glue or staple them on a sheet of paper or a moderation wall.

Look at your result. Does the supply chain make sense? What insights have you gained regarding sustainability criteria?

You have probably wondered how you can ensure that your suppliers also comply with sustainability criteria. You have certainly also wondered which parts of the supply chain you can still source regionally under the given framework conditions. You may have come across other factors and criteria such as resource-saving raw materials and production, rolling resistance of tires, transport issues and other points. This short exercise has most likely drawn you deep into the topic of sustainability in the supply chain and therefore it is now time for some theory.

We have the image of the ancient temple of sustainability in front of us and we know that we always have to consider all three pillars. Also in supply chain management we look at the three thematic categories, the economy, the social and the ecology. We have to take care of the following points if we want to ensure a sustainable supply chain:

- Sustainable supply chain → Economy
  - Securing the professional and expert knowledge within the organization
  - Creating a balance between the interests of the company and the interests of the stakeholders
  - Optimization of business processes
  - Ensuring and securing competitiveness
- Sustainable supply chain → Social
  - Impacts on society
  - Compliance with minimum standards
  - Consideration of stakeholder interests
  - Improvement of social acceptance
  - Health and safety
- Sustainable supply chain → Ecology
  - Minimum possible use of material and energy (resource efficiency)
  - Environmental and climate protection
  - Hazardous substances
  - Plant and transport safety

Do these points sound familiar to you?

Yes, because we have discussed them in detail in the last three chapters. However, it is important that we keep in mind the key points of the requirements for the supply chain when we talk about sustainability in the supply chain.

Do you remember the checklists in this chapter? With their help, we can query and ensure the most important contents from the three dimensions of sustainability for our supply chain management.

In the economic dimension, the responsibilities within the management system must be clearly regulated. This includes the commitment of the board of directors/the top management to sustainability and the self-commitment to pursue and achieve defined goals as far as possible. We then speak of a continuous improvement process. This also includes the appointment of a sustainability officer with direct reporting to the top management, as well as the fight against corruption.

In the social dimension, we have to pay attention to the compliance with social standards, as well as health protection and safety. The compliance with laws and industry standards is just as important as the consideration of the impacts of the entrepreneurial actions on the social environment, on the stakeholders. Here, regular risk and acceptance assessments are very helpful.[6]

In the ecological dimension, this involves the implementation of environmental guidelines, recycling systems and pollution prevention, as well as the exclusion of prohibited materials in products and production. We pay attention to the operational environmental and climate protection and ensure that goals and measures to reduce greenhouse gas emissions are in place. The conscious use of natural resources and the substitution of hazardous substances are in our focus (important here the labeling and storage).[7]

After we have looked at the thematic categories of a sustainable supply chain in all three dimensions of sustainability, we have to think about the prerequisites that are required for the implementation of sustainability in the supply chain. The following aspects should be considered and weighed before an implementation:

- Is an image gain to be expected?
- Is there a possible sales risk? What is the cost/benefit ratio?
- Is there a risk of failure? Does the delivery capability remain under "tightened" conditions?
- Is it possible to establish long-term and stable delivery relationships?
- Is there a quality risk? Does the quality remain? Do we find cheaper alternatives for materials and intermediate products?
- How does sustainability fit with the company philosophy? Do we need a new vision?
- Are we acting because of customer pressure or out of our own conviction?
- Which aspects of sustainability fit the company?
- How do we compare to our competitors? How can we set ourselves apart (unique selling points)?

To lighten things up, let's continue with our example from earlier: You remember, we first defined a sustainable supply chain for our car tires. Now we are (virtually) ready to present our proposal to the board/management, how we want to equip our fleet with a sustainable tire in the future, considering all steps in the supply chain, from production to recycling.

Exercise 2

- First, take the perspective of the sustainability manager and think of reasons and arguments that speak for the introduction of supply chain management in your organization. Write down the results.

- Now take the perspective of a critical board member and work out reasons and arguments that speak against the introduction of supply chain management. Write down these results as well.
- Compare your pro and con arguments and think about how you as a sustainability manager can best dispel reservations and achieve your goal in your presentation.

I have often conducted this exercise in seminars, where the participants gained valuable insights into their own organization and its viewpoints by taking different perspectives. If you prepare yourself in advance for the critical questions and killer phrases that will surely come, nothing will surprise you. Make it clear to your board in simple words that there are valid *economic* reasons for sustainability in the supply chain:

- Business risks are limited by maintaining the good reputation and brand value (branding).
- Efficiency gains can be achieved by reducing costs for raw materials, energy and transport, as well as by increasing labor productivity and the efficiency of the entire supply chain.
- The production of sustainable products and services meets the requirements of customers, consumers and business partners and provides unique selling points (so-called USPs).

Let us now take a look at the procurement processes in the context of a sustainable supply chain:

- The development of know-how is urgently required; important here are:
  - Background knowledge of the products to be purchased

- The presence or learning of the relevant language skills
- Cultural customs
- Background knowledge of critical and non-critical countries and suppliers
- Optimization of the procurement processes by:
  - Regular price negotiations with long-term suppliers
  - Supplier consolidation and bundling
  - Transparency—traceability
  - Product specification and knowledge of the required needs
  - Development of product and/or material groups

Furthermore, a supplier management should be introduced, if it does not already exist. Depending on the size and possibilities of the company, supplier managers should be responsible for certain continents, countries, country groups or regions. Through the permanent interaction with "their" suppliers, these employees are able to develop specific knowledge about the country and its people and to openly discuss problems in the daily business with their contact partners on an equal footing. This not only increases the security in purchasing, it also encourages the suppliers to openly talk about deficiencies or even ask for help. Together, a further development of sustainability can be advanced.

Communicate your sustainability expectations to the supplier. Include these expectations, especially in the form of a code of conduct, in the supplier contracts. Ask the suppliers to self-assess their sustainability performance and conduct on-site performance evaluations.

Support your suppliers in addressing existing sustainability issues. Provide them with appropriate resources, train and help them to improve their sustainability management. Provide assistance in eliminating the root causes of poor sustainability performance independently.

Also absolutely essential here is internal and external communication according to the motto: "Do good and talk about it".

- Internal communication:
  - Employees as ambassadors (credible, cost-effective, efficient)
  - Information events, e.g. open day
  - Meetings and coordination
  - Letter from the management/information brochure
  - Job interviews
  - Intranet—company magazine

- External communication:
  - Press and media work
  - Internet—homepage, Facebook, Twitter, …
  - Newsletter
  - CSR and/or sustainability report
  - Best: Linking internal and external communication (synergy effects)
  - The notice board (internal and external)

Communication is a powerful weapon for implementing SCM. With special techniques such as the use of transformative vocabulary and superlative word structures, the internal and external skeptics can be silenced. More on that later.

Cooperation with initiatives also supports sustainable management and provides new impulses for exchange with like-minded people from different sectors. Sharing best practices supports the development of one's own performance and is also part of sustainability in the supply chain.

There are many sector-specific organizations and associations that one can join, and you will either already know or find very quickly the ones that apply to you.

You should pay special attention to networking when setting up a sustainable supply chain management, because you do not have to reinvent the wheel here and can learn from the experiences of others. There are many networking platforms, such as XING and LinkedIn to name just two of them. As in any network, you have to first build a personal network and invest time accordingly. I have now built up a special network of contacts and groups in the field of sustainability over several years, from which I increasingly benefit.

Now that we have looked at the sustainability aspects in the supply chain that are the same for all organizations and companies, we want to turn to some special features.

Let us look at three important areas for the use of sustainability management: municipalities, companies and (air) transport industry. Why these areas? On the one hand, these are actors with a particularly large influence on the current and urgent issues in the context of sustainability. These include the reduction of greenhouse gas emissions, the participation and involvement of as many people as possible on the way to sustainability and the preservation and care of the natural environment. On the other hand, I can explain the mechanisms and peculiarities of a sustainability management particularly well with these groups, as I was able to (had to) make many relevant (and often painful) experiences with them. First, I would like to address the peculiarities of municipalities, as there the factor of the social dimension of sustainability becomes particularly evident.

6.7 Sustainability Management in Municipalities

The municipality on the way to sustainability

Municipal sustainability management uses methods and concepts to reduce costs and to promote sustainable economic, ecological and social development of municipalities.

The social and political expectations of the municipalities are high. At the same time, declining tax revenues, the cost pressure due to shrinking budgets and rising expenditures in the social sector significantly limit the investment power of the municipalities. The scope of action of the municipalities is shrinking, although urgent issues related to an aging society, increased expectations from politics and the population on environmental protection and increased demands due to higher living standards lead to constantly growing tasks and expenditures of the municipalities.

How can a municipality keep track of everything with the same or even decreasing staff? How are all the guidelines, urban development goals and strategic objectives to be achieved or implemented? This multitude of tasks and goals can quickly exceed the capacities of a municipal administration and be difficult to grasp in terms of content.

This is where municipal sustainability management comes in. It takes up existing community structures, links loose ends, increases the efficiency of ongoing processes and promotes the cohesion of the citizenry. Municipal sustainability management also deals with pressing issues of the present such as resource conservation, circular economy, climate protection, renewable energies, land use, biodiversity and mobility in the municipal environment.

Municipal sustainability management is a holistic approach that inevitably leads to significant cost reductions in the municipality without neglecting the interests and needs of the citizenry and the environment. It aims at a high recreational value and protection of the local natural area for the benefit of citizens and tourists. Using our three-pillar model, which covers the economic, social and ecological dimensions of sustainability, structure and methodology are given to the holistic approach of municipal sustainability management.

System boundaries

What do we actually think of when we talk about municipal sustainability management? First of all: What is a municipality?

Is a municipality identical to a community and if so, what is a community? Or are cities also a municipality, regardless of their size? Maybe size, i.e. population, would be a suitable indicator to define a municipality. I have not yet found the definitive definition for myself. Therefore, I use the political levels of a (democratic) state that are used everywhere to delimit the municipal level. Accordingly, everything below the (federal) state level is assigned to the municipal level.

Well, we have clarified the question of what. By municipality we mean all large and small communities, markets and cities. The common identifying feature is that each municipality has its own administration that controls and manages this municipality according to economic criteria, whereby by definition the well-being of the citizens always comes first.

With this we have already identified the most important peculiarity in municipal sustainability management. The citizens are at the center of municipal action.

Environment and communication

With the transport infrastructure, the energy and water supply, the wastewater and waste disposal, the municipal service offerings and shopping opportunities, the cultural offerings, the club life and other festivals and events, we have a list of responsibilities and tasks of a municipality that could be continued for a long time.

All these tasks apply to every municipality in the world, for New York City as well as for my small hometown Hohenlinden in Bavaria. All measures of the economic and ecological pillar can be applied one to one regardless of the size of the municipality. The biggest differences in the application area are in the social dimension. The communication behavior and the communication possibilities in a community with 3000 inhabitants are simply completely different from those in a district town with 30,000 inhabitants or a mega-city with 3,000,000 inhabitants. In the community, many people still know each other personally. Whether at the baker's, in the club, while walking, at the construction yard, at festivals, at the Christmas market, in the town hall, etc.—everywhere you run into each other and communicate more or less directly with each other. This is much less the case in a district town and completely impossible in the said mega-city.

For municipal sustainability management, this means that many methods and measures from the social pillar are not possible in cities as a whole. However, there is the possibility here to become more granular again and to implement in districts or neighborhoods what is simply not possible in cities as a whole. This can go so far that, for example, a neighborhood, a quarter, a high-rise building, a skyscraper forms the unit in which municipal sustainability management can be fully implemented and operated again in all three dimensions.

So we are talking about manageable numbers of people. Manageable means for me that at least 5 to 10% of the people of a certain area have communication relationships of any kind with each other. Where do I get the numbers from? Pure gut feeling, which for me as the sum of my experiences in the field of communication is coherent.

Social structures

So let us now assume that we have such a manageable quarter, neighborhood, a community, a building complex in any city, a skyscraper in a mega-city in front of us and want to successfully implement and operate sustainability management there.

In this (still) manageable area, we find structures that are well suited for the implementation of sustainable development. We often find here communal structures and human networks that can be expanded and used. Where such structures are missing, there is still often a latent willingness to participate in a process within one's own municipality, if it is sufficiently understandable and attractive.

Participating in a process that leads to a sustainable development of the municipality can be attractive. Why is that? Well, I think, each of us, who follows the trends and tendencies in important fields such as energy transition, climate change and its consequences, financial crises and the associated uncertainties of one's own savings, the connection between civilization diseases and environmental pollution more or less intensively, thinks about what needs to be done here. If we now address these people, i.e. each of us, that each municipality can contribute its part to the problem solution and that this contribution has economic and social advantages for each individual participant and that a positive contribution to the preservation of the environment can also be achieved, then we have very strong and convincing arguments at hand.

In addition, the topic has arrived at all of us. Well, let's say more or less. The confusions described at the beginning about the term sustainability have contributed to the fact that not everyone can relate to sustainability.

Communication is crucial for the success of a sustainable municipal development. Well, one could say, communication is always important. True, but here I mean the communication with as many citizens as possible.

Let's imagine, there comes such a figure in a pinstripe suit to the next citizen meeting, is briefly introduced by the administration and then starts to torture us with a 50-page presentation. This person hits us within an hour with all kinds of incomprehensible terms such as 2-degree problem, CO_2-equivalents, biodiversity, sustainable development, carbon storage system, biofuel, mitigation of the consequences of climate change and much more.

How would we react? We would leave, as fast as we can, or fall asleep after a quarter of an hour.

No, that is certainly not the right communication. Municipal sustainability management requires direct and understandable communication to and with the citizens at a central point.

So let's first discuss the type and manner of communication that is necessary to establish a successful municipal sustainability management. Much of what we discuss and fix here, we will be able to use again in the next chapter, where it is about companies.

Communication with skeptics

First of all, one thing must be completely clear to us: If we present and explain our sustainability concept in a random municipality as part of an information event, there will always be at least one skeptic (see Fig. 6.7) who will use all his or her intelligence and knowledge to prove to us and especially to his or her fellow citizens that our sustainability concept will not work.

The manifestation of the skeptic was created by the artist Thomas Fiedler in an impressive way. The skeptic stands in Bavaria, in Hörbach, municipality of Althegnenberg. The photo of the skeptic was taken by district curator Toni Drexler, who kindly allowed me to use the photo.

That this person, this skeptic (or if it gets even worse, the persons and the skeptics) will be there, that is as certain as the amen in the church. Assume that this is a secured fact and better prepare yourself for it from the outset. If you prepare yourself for this situation, you have already won half the battle. If not, if you do not prepare

Fig. 6.7 The skeptic, artist Thomas Fiedler, photo Toni Drexler

yourself thoroughly for skeptics and cannot adopt a calm state of mind, then you have lost before you even started.

Besides these predominantly resistant skeptics and general refusers, there are of course also people with fears of new things and of change processes, as well as people with false information in their heads and people who have never dealt with the topic of sustainability. For all these people, a common language must be found, if we want to be successful.

And this language must, I emphasize MUST, meet certain principles:

- Absolute honesty, no fiddling, no half-truths or embellishments
- Never a "but" on intermediate questions or objections (why, I will explain later)
- Take the viewpoint of the listeners and interlocutors
- Use transformative vocabulary (how, I will explain later)
- Speak in clear words and simple images

Trust me and take these principles as a lived, tested and successful communication style for now.

Limits

Despite all openness, empathy and diplomacy, you will sometimes encounter limits beyond which meaningful and successful work is no longer possible. I would like to illustrate this with a situation that happened to me once. Some time ago, I planned a municipal sustainability project that also included a citizen energy cooperative. The cooperative was supposed to produce as much local and renewable energy as possible, contribute to climate protection and involve the citizens on the way to a local energy transition.

A good approach at first, which has already worked many times.

I first collected the actual data of the municipality, i.e. population, number of households, business structure and of course the energy consumption for heat and electricity. Then I looked at the local conditions and potentials and quickly realized that biogas and biomass would be important topics due to the agricultural character of the municipality. Photovoltaics also promised a relatively large potential, as until then no roof of a municipal property had a PV system.

The idea was therefore to start with photovoltaics, as there are no technical surprises there and the profitability is always given with self-consumption. It was obvious to plan photovoltaic systems for the multipurpose hall and the school and to optimize them for maximum self-consumption. I do not want to go into the technical details at this point, because they were not the problem as usual. There were also considerable initial concerns to overcome, which also surprised me at first, but which every sustainability manager will face sooner or later and have to cope with.

At this point, people said things to me like "… my brother-in-law is an electrician and he said that it doesn't work …", or "… I know for sure that it doesn't pay off …", up to "… *I don't believe* that it works and even if it does, *I don't believe* that it is economical …". (We will discuss these creeds and how we should deal with them in more detail in the chapter on the sustainability manager.)

Here I had already reached a borderline point, which was not clear to me at the time. If a person argues with you with fixed creeds, you cannot oppose anything. The belief of people is the strongest and most difficult to overcome wall that you can face as a sustainability manager. Sometimes this wall is insurmountable. Sometimes, however, when it comes to belief or disbelief in technical feasibility and profitability, you can counter it with facts and sometimes succeed. This was also the case with me at first.

After I had created and explained a meticulous and detailed technical planning and profitability calculation for the photovoltaic systems, a majority was convinced.

Convinced yes, but by no means enthusiastic, there were not highly motivated potential project members in front of me, but a small group of citizens, who had doubts, uncertainty and fear of the new written in their faces despite all the facts.

However, overcoming these fundamental doubts was still the easier part. Now voices were raised that no photovoltaic system should be built on the school, because its radiation would harm the children who were taught in the school. No, this is not a joke, this was and is serious to the people there! There was a deeply rooted belief that this was the case and the fact that many private house roofs had PV systems installed, under which many children stayed, seemed strangely irrelevant for the evaluation of the school. Rather, there was a creed among many people in this municipality, which, as it turned out later, was an insurmountable limit for me.

Benefits, structure and design

If you deal intensively with municipal sustainability management, you will quickly find opportunities and potentials to initiate promising projects for cost reduction, regional marketing, strengthening local value creation and applied environmental protection with targeted citizen participation.

You then know communicative methods to explain your concept convincingly, as well as to deal constructively with objections from the municipal administration and the citizens.

How should you proceed to build structures that bring benefits to you and your municipality?

First of all, you need to build up specific knowledge and familiarize yourself with terms and interpretations on the topic. This book and other sources, especially the Internet, help you to acquire this knowledge and the terminology. To master the often so difficult step from theory to practice, you will find in appendix 4 a list for working through the fields of action for building up a municipal sustainability management. There you will find fields of action and recommendations for action for the economic pillar, as well as for the social and the ecological pillar.

You can use the fields of action for municipal sustainability management in appendix 4 as a kind of recipe. Take a few minutes and read through the recipe. Some recipes have to be cooked exactly as they are, otherwise the result will not taste good, while some recipes allow for variations, not only possible, but necessary, because not all ingredients of the original recipe are available. The same applies to the recipe for municipal sustainability management. It can and must be varied in each case, because the specific environmental components, i.e. ingredients, are different in each municipality.

Now take a pen, preferably a color marker, and mark the fields of action and measures on a printout of appendix 4 that are conceivable for your municipality. Be generous and mark also things that are still doubtful for you from your current perspective. When choosing your ingredients, make sure that there are as many ingredients from each of the three pillars as possible. This is very important, because only if a balanced choice is made here when determining the fields of action, the basis for a functioning municipal sustainability management is created. If, on the other hand, you have a clear overweight in one of the three pillars, then the dish is oversalted and will not taste good to you or your fellow campaigners.

Well, you have identified a balanced selection of fields of action and measures for your municipality. Now it is time to plan and implement your municipal sustainability management. This step is the most difficult and requires a disciplined approach, especially at the beginning. Below I have listed my recipe for the implementation of a municipal sustainability management. Here, however, I consider it mandatory that the order of the recipe steps from point (1) to point (5) is strictly followed. Otherwise, the roast is oversalted, i.e., the system does not work.

Planning and Implementation of a Municipal Sustainability Management
(1) Municipal sustainability check (fields of action)
(2) Development of the sustainability strategy for the municipality
(3) Derivation of concrete goals and measures
(4) Creating citizen participation/citizen model
(5) Implementation of measures (project management)
(6) Preparation of an annual sustainability report
(7) Regular review and adjustment of goals and measures
(8) Communication to all stakeholders

The points (6), (7) and (8) are recurring measures that must be high on the task list of the responsible sustainability managers. It is not possible to go into each of the individual points exhaustively within the scope of this book. Here I refer to the numerous and helpful literature, because my intention is to provide a guide and a meaningful structure for the development of sustainability management systems. Nevertheless, I would like to make some comments on each of the points.

The **Municipal Sustainability Check** (fields of action) should include the following topic groups, even and especially if doubts arise as to whether these topics are applicable at the municipal level at all. Believe me, there is something to be found for every topic in every municipality.

- Climate protection
- Renewable energies
- Land use
- Biodiversity
- Education
- Mobility
- Nutrition
- Cooperation with neighboring municipalities

The **Development of the Sustainability Strategy** for the municipality is indispensable for the establishment of a successful sustainability management. Where does the municipality want to be in 5, 10, 20 years?

Example: My municipality shall cover its own energy demand for electricity, heat and cold from its own, local and renewable energy sources by the year 20xx, whereby the production and distribution of energy shall be carried out by a citizen society and the added value shall remain in the municipality.

A sustainability strategy should not be one-sidedly focused on the energy topic, but rather cover as many focal points of a sustainable development as possible. These include topics such as environmentally friendly mobility, healthy nutrition and consumer-friendly regional models, demographic change (silver generation), life-long learning, sustainability as a driver of innovation, promotion of a sustainable settlement development and many more.

The **derivation of concrete goals and measures** is of particular importance. If you do not set really concrete goals at this point, whose achievement is qualitatively and quantitatively achievable and measurable, then the failure of your project is pre-programmed.

Example: We reduce the CO_2-emissions of the municipality by 30% by the year 20xx. This means that we have to produce 10,000MWh of electricity and 6000MWh of heat from renewable energy sources by then, based on our reference year 20yy. We achieve this goal with an energy quarter in the town center, which is primarily supplied by a biomass heating plant with its own combined heat and power plant (CHP), a seasonal heat storage and photovoltaic systems with electricity storage on the roofs of the surrounding municipal properties.

Creating a **citizen participation/**a **citizen model** is also indispensable. The positive and negative experiences of numerous municipalities show that citizen participation is one of the most important keys to success. It does not play a decisive role whether this citizen participation is realized through a citizen cooperative or through another construct, although there is much to be said for the citizen cooperative. The decisive factor is that a large participation of committed citizens can be achieved right from the start. For this, the use of respected citizens as multipliers and promoters is essential.

The **implementation of measures** (project management) is paradoxically one of the most difficult parts. Many people tend to discuss a complex project endlessly and go round in circles. The result is often a comprehensive, beautifully designed concept, printed on glossy paper, which then disappears into the unfathomable depths of the mayor's desk. To prevent this from happening, a professional project management is the only way to go, in my experience.

Properly set up, the project management divides the entire complex project into manageable subprojects. For each of these subprojects, clear responsibilities, target and quality specifications are defined and written down. The overall responsibility lies with the project management, which also carries out the overarching project control. A not too deep, but also not too superficial milestone planning is helpful for this. You do not have to invest much for this, because there are free software solutions for this, such as the program OpenProject[8].

The creation of an **annual sustainability report** is a great support for the implementation of measures and the project control. The report should be oriented as much as possible to the so-called GRI guidelines[9] for sustainability reporting of the GRI (Global Reporting Initiative). In Appendix 5 you will find a small argumentation aid from me, with which you can communicate the benefit of a sustainability report.

A good sustainability report is not a minor thing that you can do in a few days. In order to serve as an effective communication and control instrument, it will be more or less complex and relatively extensive. If I apply my own experience in this regard to a municipality of, for example, 2000 to 5000 citizens, then you will need about two full-time equivalents (FTE) for one year for the first-time creation of a sustainability report. Of course, you can and should distribute these FTSs to a complete team that works on the different parts of the report. Depending on the scope, you may not be able to avoid external support (data collection, data processing, editorial processing). How a sustainability report should be structured and how it is created, you will learn in Chap. 9.

The regular **review and adjustment of goals and measures,** e.g. with the help of the sustainability report, is the element that keeps your municipal sustainability management running and alive. Stick to your original goals as long as this is possible and sensible. But if you notice over time that an original goal is no longer achievable or no longer sensible, then do not waste any more resources on it. Delete this goal, end all activities associated with it and define a new, at least as ambitious goal as the original one. Do not lose sight of your strategy and your overarching goals and vary on the measure level until you notice that you are approaching your overall goal again.

The **communication with the stakeholders** (especially with the citizens) creates equal knowledge and thus probably also an equal understanding of the needs, strategies and measures of your municipality. Use your sustainability report for this purpose, which you distribute as widely as possible among your stakeholders in written and electronic form. This means, of course, above all, that you make it accessible to your citizens in a suitable form (e.g. as a PDF file on your homepage for download or partial publication in the official gazette of the municipality). Use one of the standard citizen information events, put the sustainability management of the municipality on the agenda. Report on progress but also on setbacks. Use a clear and concise form of communication, especially in your presentations (see also chap. 10).

I have summarized the strategic cornerstones for sustainable development in municipalities based on my own project experience in appendix 6.

6.8   Sustainability Management in Companies

Leading companies on the path of sustainable development is probably the most difficult thing you will encounter on the way to sustainability. What is that, you say, that can't be true. So many companies have committed themselves to sustainable development, publish sustainability reports every year, win sustainability awards. What is this nonsense?

It is not nonsense. There is a huge gap between the declarations, statements of intent and visions of many companies on the topic of sustainability and the reality or implementation in practice, which is best measured in light years, so that the places before the decimal point do not become too large.

You don't believe me?

I can't even blame you. If you read the glossy brochures, the internet appearances, the press releases and other announcements, then you have to come to the opinion that there is no need for catch-up in terms of sustainability in the companies. However, these are often enough only words without substance. By this I do not mean now that behind this there is conscious deception or cynicism of the board members. No, in my experience, these board members usually believe what is written in the said brochures of their press departments.

What is often missing there is the reference to practice and to the corresponding orders of magnitude. I once read in the sustainability report of a large German bank that the company is proud to have saved 500 t (!) CO_2 in the reporting year. Converted to the reported total energy consumption, this corresponded to approximately 0.000001 % CO_2 savings. So nothing! The board members and authors of this report were obviously not aware of the relation. Possibly they also miscalculated by several orders of magnitude and did not notice it, because no sustainability manager checked the numbers.

The biggest problem is the often missing feedback loop between the board, top management and the working level. The distance between these levels can also only be measured in light years.

I would like to make this discrepancy a bit more comprehensible to you by describing some scenes that have actually happened in a similar way.

Imagine the following situation:

Example of Missing Feedback Loop

You are a consultant, a sustainability manager and have the assignment of the board of a medium-sized company to comprehensively establish and anchor sustainability criteria and sustainability mechanisms in its operation. In the preliminary talk, the management explained the corporate goals to you, in which sustainability plays a central role. When you asked why they still need and call you, you got the somewhat vague answer that the board has recognized the importance of sustainable development and has also given ambitious goals in this direction, but that incomprehensibly the results lag significantly behind the target specifications. You are now supposed to bring the company up to speed in terms of sustainability, and the management has also told you that they have urgently asked all employees in a circular to support you to the best of their ability. In addition, they introduced you at a staff meeting to give more emphasis to their wish.

In any case, it is clear to you that the board wants your services and as a result wants to see a company that really lives sustainability. Nevertheless, you have a somewhat uneasy feeling in your stomach and that completely rightly.

This is a typical initial situation. And now we accompany our virtual consultant to the first meetings with the executives of the middle level. We start with the purchasers of the company, because there is a very important switch for or against sustainable development in the company.

We are now in the meeting room of the purchasing department. The department head is already there, probably sitting in a very dominant position in the room, his purchasers are distributed around him. The looks they throw at you are, to put it mildly, repellent. Look closely. Some of them have downright anger in their eyes. Look at their hands, they often reveal even more. Most of them have their hands crossed. Right? Right. Absolute defensive attitude.

(By the way, this scenario would probably be the same in other departments and areas of the company. You can therefore transfer this example to any area of your organization.)

Well, you are an experienced manager and you are not easily impressed. There are methods to break through this defensive attitude, and we will discuss some of them later. Just as with the skeptics, whom we first discovered in the municipal sector, we have to solve this problem and implement the visions of the board in a conflict-free way in practice.

For companies, it is helpful to address the innovation potential of the individual areas and departments. This can sometimes turn around people who were previously completely on the defensive. If you can then explain with easy-to-understand examples that sustainability does not mean being expensive, you can get controllers, buyers and other commercially oriented people on board. One danger in this case is that the people will seemingly accept your suggestions positively and then, when they are back in their office, ponder and calculate until they can "prove" to you that your suggestions simply do not pay off. "Too bad, I'm really sorry, but you see for yourself, it doesn't pay off". That's how it will come across to you, or something like that.

It has happened to me that a controller took all the risks and uncertainties that exist in any innovative project, pulled them to one side and summed them up. As a result, it turned out that the technology under investigation was 50% more expensive than the previous technology based on fossil fuel. So be prepared to be confronted with such worst-case scenarios almost reflexively. The best answer to this is to gratefully accept this worst-case scenario and supplement it with two more scenarios: best-case and normal-case.

Compare all three scenarios and you will be successful with this method, provided you are not in a completely resistant environment. If you still don't get anywhere with that, you have to do something that a project manager normally doesn't want to do and should only do in an emergency. You have to go to your client, i.e. the board or management, and make it clear to him or her that it is time for a word of power. This temporarily removes the obstacles that have held you back so far.

However, also consider the likely reaction of the people involved. They will not forget what you have "instigated" and will remember it at an opportune moment. Companies, especially large companies, tick like that, there is no way around it. A better solution for how a resistant environment can be overcome will be discussed later in the chapter on the sustainability manager.

Tip

Especially in larger and large companies, the individual departments and areas represent small 'kingdoms' that are fiercely defended against all intruders. Make it clear to those affected that you are not an attacker, but a friend and supporter. Convince them with the absolutely true argument that after the transformation to sustainability, the department and its processes will be even more successful and stable than before.

Another peculiarity in sustainability management are companies in the (air) transport sector, which are under special pressure.

6.9   Sustainability in the (Air) Transport Sector

What is the bracket in the chapter title for? I initially thought of making this chapter general. Many aspects are the same in the entire transport sector, regardless of whether we look at individual transport in the car or truck sector, whether we deal with rail or shipping traffic or deal with air traffic.

The discussion about emissions of air pollutants, greenhouse gases and noise emissions caused directly or indirectly by air traffic is very emotional and controversial. Buzzwords like "flight shame" are making the rounds and make a factual discussion extremely difficult. Therefore, this sector is ideal for exemplifying the advantages and benefits of sustainability management.

In addition, I have been working in the air transport sector for more than 30 years and therefore all the specific aspects in this area are familiar to me. I will therefore use air traffic as an example in the entire transport sector. When I now go into things like air and water pollution, noise pollution and greenhouse gas emissions, energy and fuel consumption in air traffic, just think away the part "air" and you can transfer most of it to the transport mode or modes that you have in mind.

The (air) transport sector is in a situation that cries out for a specific model of sustainability management. Although air traffic is by no means the largest among the CO_2 emitters, it is nevertheless presented and perceived by the public as the climate killer No. 1. Apparently, air traffic has been found to be the ideal scapegoat and can distract excellently from all other emitters. Because the culprit has now been named! Pure factual knowledge does not help here either, but it takes a special methodology to run an economically healthy and successful company in this sector with rising energy and fuel prices, a very negative public perception and increasingly critical and attentive customers, i.e. the passengers.

Of course, there are also other environmental impacts that (air) traffic entails. First and foremost, of course, the noise pollution, which is not only perceived as extremely disturbing, but which can also lead to health damage. The exhaust gases pollute the air with various pollutants, with the fine dust and the ultra-fine dust being a special chapter in themselves, as it can probably be considered as certain that these dusts can have a carcinogenic effect. The de-icing agents, which are applied in winter to de-ice the aircraft wings, the taxiways, the runways, the paths and roads in the airport area, sooner or later find their way into the groundwater, either wholly or partially. The degradation products from this are partly highly toxic and carcinogenic and someday we might drink such a cocktail.

The (air) transport industry is therefore crying out to be analysed and optimised from the perspective of sustainability. Only a holistic view leads to a stable economic concept in air transport, in which the ecological aspect also has a strategic value. Minimising the negative impact on the environment is almost a survival strategy for air transport. In the future, passengers will make their decision for or against a mode of transport much more than they do now, depending on sustainability criteria as well as price and speed. In turn, almost 1,000,000[10] jobs depend on these customer decisions alone in Germany. And these are jobs for unskilled, semi-skilled, trained, skilled, academic, for jobs of all kinds.

A well-thought-out sustainability management therefore also offers great opportunities for compatible future design here, whether we are talking about airports, airlines, the aircraft industry and all suppliers. However, for all companies involved in (air) transport, the following also applies: Anyone who does not seriously and convincingly devote themselves to the topic of sustainability will probably no longer exist in a few years.

The most important topic is fuel. In my estimation, the development and production of alternative and economically equivalent fuels with significantly reduced greenhouse gas emissions will be decisive for the further development in air transport.

Analysis

However, before we can discuss a possible solution, we have to look at the initial situation and the psychological component in the whole discussion pro/contra (air) transport.

Let us imagine the situation as an imaginary negotiation situation. At the meeting table, on one side of the table, sit the politicians, the consumer associations and representatives of environmental and climate protection associations, as well as a municipal representative of the neighbouring community.

On the other side of the table sits the representative of air transport and does not look very happy. The poor guy is in a permanent defensive situation and is asked from all sides to explain and justify himself. The unpleasant thing about this meeting is that no matter what the air transport representative says and what arguments he brings forward, there is ALWAYS someone at the table who has something to object.

For example, let us imagine that the representative of air transport says that the CO_2 emissions of the aircraft, the fuel consumption and the noise emissions are constantly decreasing and are already x percent lower than y years ago. Immediately, the gentlemen of the environmental associations jump up and shout excitedly "... everyone knows that aircraft are the climate killer No. 1 ...". It does not help the gentleman from air transport at all in this situation that a) the statement is wrong and b) that the politician at the table is actually on his side and ready to acknowledge the successes of the reduction measures.

It does not help, because the mood has immediately become very emotional and our politician does not want to spoil it with the environmental and climate protection associations (he wants to be re-elected). If the transport representative then states that air transport creates and maintains many jobs, which is again supported by the politician, the gentleman from the environmental association immediately says that this goes hand in hand with large-scale sealing of land, and the representative of the surrounding area adds that the housing costs and rents are constantly rising because of the settlement of air transport industry, airports and their suppliers and poorer sections of the population are therefore moving away.

I could play this imaginary negotiation situation for a long time and it would not become more constructive. As long as the air transport representative tries desperately to convince with pure factual knowledge, he will fail, because there are also enough facts that can be used against him and his industry. This fight cannot be won like this. Just as there is a perceived coldness for people, there is also the perceived knowledge that (air) transport is per se harmful and dangerous and that one should be ashamed of every flight one takes. We have to understand and accept this if we want to develop a successful strategy against it.

In many discussions with environmental and climate protectors (I count myself among this species) I have deliberately and persistently asked what the opponents of air transport base their opinion and feeling on. The first sentence that comes up is: "Everyone knows that" and it is very hard to fight against that. The second is: "I just have to look at the sky, then I can see the dirt of the planes" (meaning the contrails that form behind the engines). And the third is: "Don't you hear how insanely loud that is?". The countless vehicles that blow exhaust gases into the air and cause noise every day around us are not perceived, because they are simply too many and as self-evident in everyday life as our mobile phones. We normally do not notice pollution from power plants and industrial plants at all, unless we live near them.

Moreover, it is so convenient to condemn air traffic as the number one climate killer. It is so visible and distracts so wonderfully from all other emitters. "Let us stop air traffic and we will live in a happy world" is the credo of most opponents of air traffic, who nevertheless want to fly on holiday two or three times a year. They feel a little ashamed and perhaps atone for their guilt by buying an indulgence, i.e. a compensation certificate. It is schizophrenic, yet we have to deal with this very closed block if we want to carry out successful sustainability management in the (air) transport sector.

Strategic approach

For successful sustainability management in (air) transport, we also have to remember the basic principle of sustainability here—the equal consideration and use of the economic, social and ecological dimension.

For the economic approach, we have very good arguments such as value creation, jobs, growth and much more. Also in the ecological field, we can score with many positive approaches and concepts. The constant reduction of fuel consumption and emissions are impressive arguments that cannot be denied. However, we cannot convince with this, as we have already established.

The decisive factor is the social dimension of sustainability. Here we have to intensify our efforts and catch up on what was largely missed in the past. We have to convince people emotionally, with the aim that air traffic as a whole is perceived and felt as good. For this, a comprehensive communication concept must be developed and disseminated as far as possible.

This is not about propaganda, as some might now feel, but about presenting and presenting factual knowledge in such a way that a positive access is possible. The use of transformative vocabulary can be very helpful here.

Let us take as an example the new generation of engines with the gearbox in the engine (Geared Turbofan—GTF). We do not need to go into the details now, which are connected with the decoupling of the fan and the low-pressure turbine. For our example, the fact that a GTF engine is much quieter (by about half) and consumes much less fuel (by about 15 to 20%) than conventional engines is sufficient.

Now let us take a look at how this, yes one has to say, groundbreaking news was presented in a newspaper article some time ago. Read it through:

An Objective Example

Using the example of Sample-Airport airport, this means, Mr. John Doe explains, that a plane landing from the east does not spread its noise carpet around the whole Sample-City. Only shortly before the municipality of Sample-City, 75

> decibels (dB) are reached. With the conventional engines, it is already 80 dB and more. Mr. John Doe finds it "not contemptible" that the new drive system requires 15 to 20% less fuel at the same time.

The technical data and the spread of the noise carpet are quite correctly presented, but are you enthusiastic about this article? Do you already sense what this new engine will bring, for you, for the environment, for the airport surroundings? No.

How about a version like this:

An Inspiring Example

Using the example of Sample-Airport airport, this means, Mr. John Doe explains, that a plane landing from the east will hardly be heard in Sample-City, as the new engines are half as loud as conventional engines. Only shortly before the municipality of Sample-City, you will hear the plane again consciously. Mr. John Doe finds it "fantastic" that the new drive system requires 15 to 20 % less fuel at the same time, because this means not only less costs for the airline, but above all less air pollutants and less greenhouse gas emissions and that is good for the environment!

I think that this version will appeal to you much more strongly. Again, this is not about propaganda. When I speak for a human and emotional communication, then from my own experience, how we humans want to be addressed, if we want to achieve a change in behavior. And that is the point and the joke of the matter in this small example. The person who gave the interview for this article probably had only the best intentions and wanted to contribute through their statements to the fact that this new technology is increasingly used. But they did not! I followed the appearance of this article at that time and the public reaction to it. Reaction equal to zero!

And that is exactly where we as sustainability managers have to start. If we have recognized a positive development and helpful technical innovations and want to implement them, we have to address the people we want to reach vividly and emotionally. If this had been done, as shown in my small example, then it might have been possible to create a positive expectation in the public and to build up the necessary pressure that the Sample-Airport airport would have offered reduced landing and take-off fees for aircraft with this "whisper engine". This in turn would have led to a displacement process in favor of the new, quieter and less emission-intensive aircraft.

Similarly, of course, this would also work for all other modes of transport. It is crucial that we sustainability managers understand ourselves more as "sellers" of good ideas and technologies and communicate accordingly enthusiastically.

Besides the right communication, it is also about optimizing the processes in (air) traffic and thus achieving an even higher energy and fuel efficiency. For example, the global introduction and application of A-CDM (Airport Collaborative Decision Making) would be an important step in this direction. With A-CDM, all processes of a flight in the air and on the ground are analyzed and optimized with all stakeholders. This benefits both the airlines with lower fuel consumption and the airport operators with a more efficient and cost-saving handling by ground traffic and terminal services, and also the passenger benefits from shorter waiting times. It would go too far to explain the A-CDM system in detail here. Anyone who wants to deal with it more intensively, I recommend the page http://www.euro-cdm.org/. In any case, a truly sustainable process that also needs to be communicated accordingly.

Very often, the technical and organizational problems with new technologies and processes are already solved and are still not widely used. Here we sustainability managers are in demand and we have to understand ourselves as "implementation project managers" who transfer these positive things, techniques, innovations into common practice. In this way, we create sustainable improvements for people and the environment and can communicate these positive developments without hesitation.

Another very important approach is the fact that (air) traffic enables very fast connections. This sounds very trivial at first, but think about it for a few minutes. If there was no (air) traffic, how long would it take you to get to your holiday destination, to the branches of your company all over the world, to attend meetings and conferences that require personal presence? Days and weeks, instead of hours! This, already taken for granted fact, must be worked out and emphasized and this in a way that appeals to every human being emotionally immediately.

The discussed peculiarities of sustainability management for municipalities, companies and in (air) traffic are sectors in which I want to illustrate the challenges of sustainability management. I have taken them, as I said, because I know them and also wanted to illustrate with examples that we have to consider not only generally valid concepts and methods, but also the peculiarities and specific conditions of each organization and sector. When we later deal with the key to success in the book, we will look at the topic of peculiarities of organizations from another perspective.

A very important tool, how we sustainability managers can lead an organization of any kind and independent of the sector towards sustainability, is the methodical review of this organization with regard to the compliance with sustainability criteria.

6.10 Certification of the Sustainability of Organizations

The idea of creating a certification system to prove the sustainability of organizations came to me after many considerations on the question of how to best inspire a hesitant management of the many advantages of a primarily sustainable and not only classically economically managed organization. The result of my considerations was then the realization that this can succeed if sustainability becomes measurable. For this purpose, key figures—also called KPI[1]—are used. Ideally, the degree of sustainability determined in this way is certified by a neutral organization and the audited company then receives the corresponding quality seal.

There are quality seals like sand on the sea and many of them are at least questionable. Nevertheless, most people trust a product with a quality seal more than one without. This effect should be taken into account.

In addition, the time is simply ripe to make the switch from a primarily economically managed company to a primarily sustainably managed company. Also, organizations and their stakeholders increasingly recognize the necessity and the advantages of socially responsible, sustainable action. This applies especially to energy-intensive and resource-consuming companies, which on the one hand

---

[1] KPI: Key Performance Indicator, a key figure that can be used to measure or determine the progress or degree of fulfillment with regard to important objectives or critical success factors within an organization (source: Wikipedia).

fulfill the desire of billions of people for consumer goods of all kinds, on the other hand, however, face fierce criticism in the public. Noise, air and water pollution, climate-damaging emissions and soil sealing are just some of the numerous fields of criticism that every organization that produces products or offers services has to deal with.

In order to change the public perception in a positive sense, to reconcile these organizations with the environment and the people, the assumption of social responsibility through sustainable action is essential for the affected organizations. This is becoming increasingly clear to managers all over the world.

A quality seal for sustainability confirms that the certified organization has implemented comprehensive sustainability criteria within its own organization and also lives up to them. The audit contents of such a quality seal should be oriented towards the ISO 26000 "Guidance on social responsibility" and adopt recommendations from it that are adapted for the quality seal and the goal of a sustainable organization. Then this quality seal contributes to the fact that certified organizations can achieve significant improvements in the economic, social and ecological dimension of their activities. These are mainly:

- Increase of competitiveness
- Enhancement of reputation
- Increase of the ability to attract or retain staff or members
- Increase of the ability to attract or retain customers and clients or users
- Increase the positive assessment and evaluation by investors, owners, donors, sponsors and the financial world

- Improve their relationship with businesses, governments, the media, suppliers, partners, customers and the community in which they operate

The contents required for certification should be recorded by means of a checklist[12] and appropriate evidence should be requested for the provision of the queried contents. From the ISO 26000, core topics should be included that I consider essential for the assessment of sustainability in organizations:

- Environment
- Consumer concerns

Also, the procedures should be included:

- Recognition of social responsibility
- Identification and involvement of stakeholders
- Communication on social responsibility
- Improvement of credibility in the context of social responsibility

In addition, I recommend to include methods and procedures that can positively change the public perception especially in the manufacturing and transport sector:

- CO_2-Footprint within the system boundaries of the IPCC[13] (Scope 1–3)
- CO_2-Reduction measures for energy and fuel consumption
- Sustainability reporting, sustainability strategy and management

The certification process could then look like this:

- The certification of the sustainability of a company or organization is based on a checklist filled out by the company and the provision/submission of corresponding evidence.
- The completed checklist is then checked by the respective auditor for completeness and plausibility.
- Based on the information and evidence, the auditor selects important contents for the assessment, which are then examined on site in an on-site audit. This on-site audit is very important, so that the auditor gets an accurate picture of the organization to be audited and can approach the next step confidently.
- Based on a generally valid evaluation system and the submitted documents, as well as based on the findings from the on-site audit, the quality seal is then awarded if the minimum qualification is achieved.
- After the audit and evaluation, a certification report should be prepared and explained to the top management of the company. The certification certificate can then be presented in a public-effective manner.

I have placed great value in my system *Sustainability. Now.*[11] that the criteria that lead to obtaining this certificate are demanding and yet achievable. A greenwashing is certainly avoided if the auditor strictly adheres to the evaluation system.

As I said, this is not about deception. It is rather about capturing and communicating the achievements of an organization in the field of sustainability. The motto is: Do good and talk about it!

I have often experienced that already in this phase of data collection the participants view their operation from a completely new perspective and find astonishing ideas and approaches on how their organization can be optimized. The auditor does not have to do much in this phase. In this phase it is more a moderation than a consultation according to the motto: guide and influence.

Based on the collected data and the important on-site audit, statements can then be made about the degree of sustainability in the organization. This is followed by recommendations and suggestions on how the organization should develop in terms of sustainability until the next review (suggestion: every 3 years). There are some things that I consider indispensable and that every organization can quickly and easily establish:

- Sustainability report (see chap. 9)
- Sustainability strategy across all three dimensions and including sustainability management
- Sustainability officer, reporting directly to the top management
- Goal and action planning to optimize the organization in all three dimensions of sustainability
- Anchoring the measures in the target system of the executives (if not available: introduce target system!)

Of course, you do not need to award your own quality seal to examine and optimize an organization on its degree of sustainability. However, you do need a checklist that covers all three dimensions of sustainability in any case. I think, after reading this book and especially with the tools in your backpack, you are equipped to create such a checklist. The checklist "Mitigation of Climate Change" in Appendix 8 can serve as a model for you.

Conclusion: The certification of lived sustainability in organizations is a very good method to anchor sustainability. For this, however, the degree of sustainability

must be made measurable. This is achieved with suitable indicators.

Source Reference and Notes

1. Energy can neither be generated nor consumed, but only converted (2nd law of thermodynamics). Colloquially, I use the word energy consumption, because the converted energy (usually heat) is no longer usable for us, i.e. "consumed"
2. See http://wirtschaftslexikon.gabler.de/Archiv/21339690/oekologische-nachhaltigkeit-v2.html
3. see World Resource Institute, Greenhouse Gas Protocol (GHG)
4. Guidance on social responsibility (https://www.din.de/de/mitwirken/normenausschuesse/naorg/veroeffentlichungen/wdc-beuth:din21:330481644)
5. See https://www.globalcompact.de/wAssets/docs/Weitere-Themen/hintergrundpapier_innovation_und_nachhaltigkeit.pdf
6. See also ISO 26000, http://www.iso.org/iso/home/standards/iso26000.htm
7. see also ISO 14001, www.iso.org and EMAS, www.emas.de
8. https://www.openproject.org
9. https://www.globalreporting.org
10. Source Federal Association of the German Air Transport Industry (BDL, www.bdl.aero)
11. More information on this at: https://www.nachhaltigkeit-management.de/
12. see appendix 8, example of a checklist for assessing sustainability
13. IPCC stands for: Intergovernmental Panel on Climate Change

7

Making Sustainability Measurable

A functioning sustainability management is based on reliable data from all three dimensions of sustainability and their relation to important sustainability metrics in the form of indicators and criteria.

To determine the progress or the degree of fulfillment in the implementation of sustainability criteria, a key performance indicator (KPI) system for the organization should always be established. For companies, KPIs are defined that are either industry-specific or specific for the company or the organization. But which KPIs are relevant? This question cannot be answered generally. For each company and organization, a separate selection has to be made. The definitions of the GRI standards[1] can help. Although they were developed for sustainability reporting,

---

[1] GRI = Global Reporting Initiative, defines guidelines for the preparation of sustainability reports, the GRI Standards

© The Author(s), under exclusive license to Springer-Verlag GmbH, DE, part of Springer Nature 2023
M. Wühle, *Making Sustainability Measurable*,
https://doi.org/10.1007/978-3-662-66715-6_7

parts of the system can be used to develop **material sustainability indicators** in the company.

The materiality is defined by topics—the Material Topics—as well as by the impacts and how the company deals with these impacts. The Material Topics provide a balanced picture of the essential topics of the company. The Impacts describe the effects of the company on the economy, the environment and the society.

My recommendation at this point is: Conduct a materiality analysis according to the GRI standards. The materialities and the associated KPIs that are relevant for you, your company, your organization will be clear to you afterwards. For a producing company, for example, the following materialities could emerge:

- Product safety
- Procurement and supplier management
- Sustainable product development
- Efficient use of natural resources
- Climate protection
- Circular economy

From this, the material KPIs for your company or organization can be developed. There are many sources and access paths to determine the material KPIs and there is nothing wrong in principle with developing them for the company itself. However, the wheel does not have to be reinvented here either. The German Association for Financial Analysis and Asset Management (DVFA) has developed key criteria for the most important sectors (sectors) under the name *KPIs for ESG*[2]. A selection of these

---

[2] KPIs for ESG 3.0 (Key Performance Indicators for Environmental Social & Governance Issues) define criteria with one or two performance indicators for 114 subsectors according to Stoxx Industry Classification Benchmarks. This reporting standard has received great resonance since its publication, and is still considered the standard work, although published in 2008.

can be used well for your indicator system in the first step, based on experience.

An excerpt from the sector 2757—Industrial Machinery as an example shows the basic structure of the DVFA system for KPIs:

- *Energy Efficiency:* Energy consumption total
- *GHG Emissions:* GHG Emissions total
- *Innovation:* Total R&D expenses
- *Emissions to Air:* Total CO_2, NOx, SOx, VOC emissions
- *Eco-Design:* Improvement rate of product energy efficiency compared to previous year/product
- *Supply Chain:* Total number of suppliers
- …

The materiality analysis is basically a learning process that is subject to a regulated adjustment mechanism. Data is continuously collected from which the relevant KPIs (key figures) are formed (See Fig. 7.1). These KPIs can of course also be used for the sustainability report of the organization, as they make the degree of sustainability of the product or the organization measurable with their help and interpretation. **The primary goal is always to be able to better control the transformation process towards sustainability.**

In the entire process of implementing sustainability criteria and sustainability goals in an organization or a company, two questions always arise:

- What change/improvement should be achieved?
- How can this be measured?

At the beginning, sustainability goals must be defined that are ambitious but also feasible. The selection of the KPIs

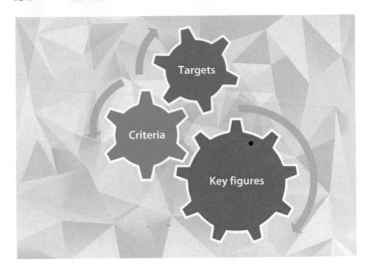

Fig. 7.1 Indicators to make sustainability measurable

that correspond to these goals are compiled anew each time.

This makes the sustainability process a continuous improvement process that leads to quickly visible successes in all three dimensions:

- *Economic dimension:* Reduction of costs (energy, raw materials, time), crisis-proof supply chain
- *Ecological dimension:* Less pollutants, reduction of resource consumption and CO_{2eq} emissions
- *Social dimension:* Increase of customer satisfaction, reduction of employee turnover

Helpful for the definition of the KPIs is the system of 'scopes' from the Greenhouse-Gas-Protocol. In this system, the scopes serve to assign the responsibility of CO_{2eq} emissions. However, the underlying logic of the associated system boundaries can be very well used for all KPIs.

- *Scope 1: KPIs that lie directly and immediately in one's own area of responsibility* e.g. energy consumption and resource consumption in development and production, partly also of the later product
- *Scope 2: Energy-related KPIs in the indirect area of responsibility* e.g. purchased electrical energy, steam, heat and cold
- *Scope 3: KPIs that lie primarily in the area of responsibility of the suppliers, but can be indirectly controlled* (sustainable supply chain) e.g. energy and resource consumption in the production of intermediate products and purchased materials, transport, waste
- *Scope 4: KPIs of the suppliers in the indirect area of responsibility*e.g. energy and resource consumption, waste and recycling rate in the entire supply chain

The resulting sharper delineation regarding direct and indirect responsibility for the current value of the corresponding KPI shows who needs to be contacted in order to achieve a desired target value.

An advantage of this system is also that for a sustainability report, an interface is already available that can be used for the creation of the CO_2 footprint without further adjustments.

KPI categories

Before we can now compile and define the KPIs for our company, some basic considerations on KPIs in general have to be made.

The selection of the KPIs should not be based on general popularity, but derived from the sustainability strategy and the specific characteristics of the project. KPIs should be selected from all dimensions of sustainability as evenly as possible, so that a balanced picture of the company's degree of sustainability emerges.

Definition of the goals and determination of the KPIs:

- Are the given/defined goals in terms of sustainability for our organization ambitious and feasible?
- If not, what needs to happen to make them ambitious and feasible (e.g. conducting a feasibility study)?
- What indicators are required to make the achievement of the goals measurable?
- Are the chosen KPIs balanced across all dimensions (economy, ecology, social)?
- Are both quantitative indicators (e.g. saving of CO_{2eq} emissions per year) and qualitative indicators (e.g. customer satisfaction) present?
- Which KPIs are essential for the project, for the organization?

Once these questions are answered, the indicators are selected and target values are defined for each indicator, the last step is taken. The KPIs are assigned to certain categories that allow a more precise evaluation of the different influencing factors. In addition to the focus area, i.e. the primary assignment of a KPI to one of the dimensions of sustainability, there are four categories to which we assign our developed indicators.

First of all, there are **input-KPI,** whose value cannot be improved or optimized by ourselves, but can only be changed by appropriate negotiations and agreements with the supplier. This would be, for example, the CO_{2e} emission of the precursor or product that we purchase.

In the process itself, the **control-KPI** (e.g. waste or hazardous substances) and the **disturbance-KPI** (e.g. employee turnover) can be directly controlled, or their influence on the process or the company can be at least reduced by measures.

The **output-KPI** provide information about the quality and the profitability of the products or services. The output is distinguished between efficiency and effectiveness-KPI. For example, the production and material costs are typical **efficiency-KPI** and the complaint rate is an example of an **effectiveness-KPI.**

Now we have developed all the relevant and essential indicators for our company. The assignment to different system boundaries (scopes) and categories result in a KPI-catalog for our company, with which we can determine the degree of sustainability and optimally control the sustainability process. We need the KPIs at many points. For the creation of a life cycle assessment, or a CO_2 footprint and as data material for the sustainability report. The sustainability manager thus gets a tool at hand that is indispensable for the management of sustainability. We will deal with these topics in the next chapters.

8

Life Cycle Assessment and CO_2 Footprint—Two Sides of the Same Coin?

A life cycle assessment considers all environmental impacts caused by a company, a municipality, an organization of any kind. The CO_2 footprint (Carbon Footprint) considers a subset of these, the CO_2 emissions (or CO_2 equivalents, CO_{2eq}), which are directly or indirectly caused by an organization. Both life cycle assessment and CO_2 footprint can be applied not only to companies or organizations, but also to products and services.

8.1 Life Cycle Assessment

Life cycle assessment is a method to capture and evaluate environmentally relevant processes. Originally developed mainly for the evaluation of products, it is now also applied to processes, services and behaviors. From the perspective of sustainability, a life cycle assessment is

© The Author(s), under exclusive license to Springer-Verlag GmbH, DE, part of Springer Nature 2023
M. Wühle, *Making Sustainability Measurable*,
https://doi.org/10.1007/978-3-662-66715-6_8

ideal when a closed cycle is achieved. This cycle reflects the highest possible resource efficiency.

For the creation of life cycle assessments, the compliance with two principles is important. On the one hand, the cross-media approach. It covers all relevant, potentially harmful effects on the environmental media soil, air and water by the company under consideration. On the other hand, we have the so-called material flow-integrated approach. This refers to all material flows that are associated with the system under consideration (e.g. raw material inputs and emissions from upstream and downstream processes, from energy generation, from transports and other processes).

> **Definition**
>
> According to the rules of the standard EN 14040, a life cycle assessment comprises the definition of the goal and scope, an inventory analysis, an impact assessment and an evaluation of the collected data. After defining the goals and scope of the life cycle assessment, the essential elements are then captured and described.

It starts with the so-called inventory analysis of the material and energy flows over the chosen life cycle. Within the company-specific system boundaries, it captures and balances the input and output variables.

Next, the construction phase is optionally described. What environmentally relevant processes are there during transport to the construction site and during installation of systems and equipment in the building? This is followed by the also optional use phase. How does the use, application, maintenance, replacement and renewal of the product/system/equipment, as well as the energy use and water consumption, proceed?

Optionally, the disposal phase is now captured. What environmentally relevant things happen during dismantling or demolition and during transport to waste treatment? How is the reuse, recovery or recycling or disposal of the product, system, equipment carried out? (Fig. 8.1)

Then it is described and captured what goes into and out of the product, system, equipment. The input in terms of energy, water, raw material, intermediate products, land use, as well as further information such as compressed air, fuels or auxiliary materials. The output in terms of waste heat, emissions to air, water and soil, waste and generated products and by-products is captured. Waste must be

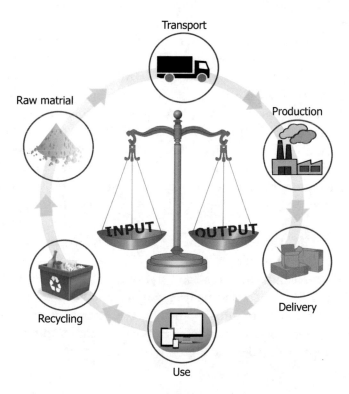

Fig. 8.1   Life cycle assessment and circular economy

subdivided into hazardous waste for disposal, non-hazardous waste for landfill and radioactive waste.

Impact Assessment
Once the inventory is completed, the impact assessment follows. Here, the size and significance of potential environmental impacts of a product system over the course of the life cycle phases are identified and assessed. This is done by referring to the inputs and outputs of the inventory analysis.

Now the evaluation takes place. Here, significant parameters of the life cycle assessment are described and assessed, conclusions are drawn and recommendations are made.

All points of the life cycle assessment are summarized in a report. It describes in detail and as comprehensibly as possible the goals, the scope of the investigation, the inventory analysis, as well as the impact assessment. The report also contains the necessary conclusions from the results of the inventory. With the help of the report, the affected organization recognizes existing and future environmental impacts from its business activity. It can then take appropriate measures to avoid or compensate for them.

The Life Cycle Assessment is a Prerequisite for a Circular Economy
What recommendations and measures can be derived from a life cycle assessment? In any case, a concrete catalog of measures to improve the life cycle assessment should be created, together with a corresponding communication concept to all stakeholder groups of the company.

The life cycle assessment is the basis for assessing the eco-efficiency of the affected organization and its products. Even though many managers and executives are still skeptical about the topic, there are undoubtedly many advantages that can arise for a company from eco-efficiency projects. A higher eco-efficiency usually also leads to an efficiency increase of the production. The reduction of raw material and energy use in turn lowers the production costs. The creation of a life cycle assessment is the ideal occasion for training and further education of employees in the areas of environment, production and costs. In addition, synergy effects often arise from cooperation with other companies and organizations that are going the same way. The introduction of a functioning circular economy in the company is thus relatively easy.

8.2   CO$_2$-Footprint

Generally, a distinction is made here between the CO$_2$-footprint for products (PCF: Product Carbon Footprint) and the one for companies (CCF: Corporate Carbon Footprint).

The CO$_2$ footprint or the CO$_2$ balance is the total measure of CO$_2$ emissions and/or greenhouse gas emissions (GHG) in CO$_2$ equivalents (CO$_2$eq). The CO$_2$ footprint is a useful tool to determine the climate impacts of products, services and organisations.

The CO$_2$ footprint of a product refers to the balance of GHG emissions along the entire life cycle of a product (cradle-to-cradle). The product cycle includes the entire value chain. From the production, extraction and transport of raw materials and intermediate products to the production and distribution to the use and possible reuse.

The end of the product cycle is the disposal or recycling of the product.

The CO_2 footprint of an organisation is often created for the first time for the sustainability report. The corporate carbon footprint covers the total $CO_{2\text{-eq}}$ output within the specific system boundaries of the company, the IPCC system boundaries from the "Greenhouse Gas Protocol" (http://www.ghgprotocol.org/).

Companies that have embarked on the path towards sustainability quickly recognise that determining their own CO_2 footprint is an important and indispensable starting point for all further steps of sustainability management. For the creation of a life cycle assessment, it is one of the most important indicators (Fig. 8.2).

Scopes—System Boundaries of the Company
The system boundaries are of particular importance. We have already encountered them in the previous chapter on key figures. Where does the responsibility of the respective

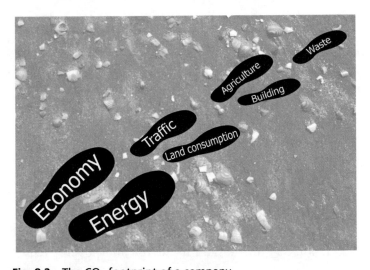

Fig. 8.2 The CO_2 footprint of a company

organisation for greenhouse gas emissions begin and end? The Greenhouse Gas Protocol provides a definition that makes it relatively easy to identify the responsibilities. The so-called "scopes" are used to divide the greenhouse gas emissions into directly and indirectly responsible emissions.

When creating the CO_2 footprint, it is advisable to structure the emissions within the company's area of responsibility according to the scopes. Scope 1 covers all direct greenhouse gas emissions caused by the organisation and Scope 2 covers all indirect GHG emissions. Afterwards, Scope 3 can optionally be recorded, all indirect GHG that are not directly responsible, but indirectly caused by the organisation.

Direct or Indirect Responsibility?
This sounds a bit theoretical, doesn't it? Therefore, the following example for a typical medium-sized company should illustrate the scopes:

- the own combined heat and power plant (CHP) and fleet of the company: *Scope 1*
- the purchased electricity, the purchased heat from an energy supply company: *Scope 2*
- the necessary infrastructure, e.g. motorway access road that was built for the organisation, emissions of service providers, waste disposal, product use, outsourced activities: *Scope 3*

To create the CO_2 footprint, the first step is to divide it into scopes. Then the corresponding GHG balance is created, i.e. the calculation of the corresponding emissions in CO_2 equivalents (CO_{2eq}).

The necessary energy values and fuel values must first be determined for the affected company, which can be more or less time-consuming the first time. The emission factors, on the other hand, can easily be taken from the annual accounts of the energy suppliers, which must be shown there in many countries. There are also other lists of national and international emission factors on the Internet, including on the website of the Federal Environment Agency. The emission factors change annually and reflect the successes and failures on the way to the energy transition.

However, the effort to create the CO_2 footprint should not discourage. Together with the life cycle assessment, the CO_2 footprint enables the establishment of an efficient sustainability management and a circular economy in the company and is an indispensable basis for every sustainability report.

Life Cycle Assessment and CO_2 Footprint—Two Sides of a Coin

Are life cycle assessment and CO_2 footprint really two sides of a coin?

In principle yes, although the life cycle assessment focuses on the environmental impacts and the CO_2 footprint clearly targets the climate effect. However, both sides of the coin have in common that they show a way towards sustainability. A way that every company must take that wants to transform itself into a sustainable organization. A life cycle assessment with a corresponding CO_2 footprint of the affected organization are two aspects of a common and successful path for companies to sustainability.

The Right Approach

But which of these aspects—life cycle assessment or CO_2 footprint—is the right approach for which company?

The question cannot be answered like that, there are too many framework conditions in each company that have to be considered and evaluated. However, it has been shown that for energy-intensive companies, for example in mechanical engineering, the CO$_2$ footprint represents the easier starting point, as GHG emissions and energy consumption are often directly proportional here. For producing companies with a large supply chain and numerous intermediate products, on the other hand, the life cycle assessment is often the better approach, as resource consumption is the determining factor here.

Finally, it does not matter how a company starts here. The creation of a CO$_2$ footprint ultimately also always leads to a resource or life cycle assessment. And when creating a life cycle assessment, the topic of energy consumption and thus also GHG emissions is always an essential component. In this respect, they are really two sides of a coin and therefore it does not matter which of the two methods the start is made with.

Both sides of the coin are important components at the beginning of every sustainability management. A sustainability management that, as a very important control instrument for every company, makes the transformation to sustainability and to a circular economy possible in the first place.

Finally, the life cycle assessment leads to the CO$_2$ footprint and vice versa. Purely a matter of taste.

Conclusion

A life cycle assessment and a CO$_2$ footprint are two aspects of the same thing. They are both necessary to transform a company towards sustainability. Both sides of the coin lead to a functioning circular economy.

Source Reference and Notes

1. http://www.ghgprotocol.org/
2. here the perspective is meant, from which we look at the different emission sources
3. Specific emission factors for the German electricity mix: https://www.umweltbundesamt.de/themen/luft/emissionen-von-luftschadstoffen/spezifische-emissionsfaktoren-fuer-den-deutschen

9

Sustainability Report— Compulsory or Freestyle?

Regardless of which form of organization we prefer to implement sustainability, we have to talk about what we are implementing and communicate our results. But why should an organization create a special sustainability report for this purpose? It is enough to convert the organization as described to the application of sustainability criteria, or not? We are convinced that a sustainable acting organization is also a successful and stable organization. So why report on it, wasting unnecessary time and resources?

The answer is that the sustainability report is the "kit" that ensures that the organization does not slacken in its efforts regarding sustainability over the years. Because with the sustainability report, both the successes and the challenges for the organization become visible on a regular basis. Both create the necessary pressure on the decision-making level within the organization. Once a sustainability report has been published, the management will not abolish it again the next year. The stakeholders of the company will ensure that the reports continue to be produced.

For sustainability management, the report is the basis par excellence, on which the goals, projects and measures are defined and monitored. In terms of external impact, the sustainability report also contributes significantly to the image of the organization and can help to smooth the waves in difficult industries.

I would even go so far as to claim that a sustainability report is an indispensable measure for any organization that wants to be permanently successful in a few years' time. The sustainability report helps the organization to set goals, measure performance and create change, in order to be able to design its own business activity sustainably.

This is vital, because the previous, purely profit-oriented business philosophy has become obsolete. Consumers, customers and suppliers are increasingly oriented towards sustainably acting organizations and make this not least also dependent on the existence of a regular sustainability report.

It can sometimes be hard to convince the management, the boards of directors of the necessity and the many benefits and opportunities of a sustainability report. For this case, I have included a small benefit list in appendix 5, which you can use for your persuasion work.

The sustainability report is a central element of any sustainability management. That is why we treat this point relatively extensively.

9.1   Procedure and Structure

For the preparation of a sustainability report, there is an international "quasi-standard"—the "Guidelines for Sustainability Reporting" of the Global Reporting

Initiative (GRI)[1]. There is no obligation to use these guidelines, but I recommend that you at least follow them when you create your first sustainability report. If you create your report "in accordance" with the GRI guidelines, you are obliged to inform the GRI about it after publication. The same applies if your report contains standard disclosures from the guidelines, but does not meet all the requirements of the "in accordance" option. You can provide your sustainability report to the GRI in written or electronic form, as well as register it in the GRI's online database[2].

For the sake of simplicity, I assume in the following that you want to create your sustainability report "in accordance" with the GRI guidelines. This raises the question of whether you choose the option "in accordance with the core" or "in accordance with the comprehensive" for yourself. The "core" option contains, in my opinion, all the essential elements and disclosures of a sustainability report. The "comprehensive" option builds on the "core" option and expands it with additional disclosures on strategy and analysis, corporate governance, as well as ethics and integrity of the organization.

In case you decide to choose the "comprehensive" option, you will not be spared from dealing intensively with the GRI guidelines[3]. The corresponding scope would go beyond the scope of this book, so I would like to refer to the literature on sustainability reporting, as well as to the GRI guidelines themselves. A comprehensive collection of articles on sustainability reporting can be found, for example, in the 'Lexicon of Sustainability'[4].

I will limit myself here to the "core" option, which is completely sufficient if you are dealing with the topic for the first time. Also here, of course, my recommendation to inform yourself further if possible and time resources allow. A good introduction is also provided by the

"Recommendations for Good Corporate Governance"[5] of the German Federal Ministry for the Environment, Nature Conservation, Building and Nuclear Safety.

What disclosures must a sustainability report in accordance with the GRI guidelines "in the core" contain? These are:

- General standard disclosures
- Specific standard disclosures (management approaches, the Disclosures of Management Approach, abbreviated DMA and indicators)

The General Standard Disclosures include the following chapters and topics:

- Strategy and analysis
- Organizational profile
- Identified material aspects and boundaries
- Stakeholder engagement
- Report profile
- Corporate governance
- Ethics and integrity
- Sector-specific general standard disclosures

The required specific standard disclosures and indicators are:

- General disclosures on the management approach
- Indicators
- Sector-specific specific standard disclosures

If parts of these disclosures are already contained in other reports of your organization, such as the annual or the environmental report, you do not have to repeat these contents in detail, but it is sufficient if you indicate in your report where these information can be found.

9.2 Basic Principles of Content and Quality

First, you should be aware of the principles that are required for a sustainability report in terms of content and quality according to the GRI guidelines:

- **Stakeholder engagement**
  Identify your stakeholders and explain how you have addressed their reasonable expectations and interests.
- **Sustainability context**
  Present the performance of your organization in relation to your concepts of sustainable development in all three dimensions of sustainability (economy, society/ social, ecology).
- **Materiality**
  Report the material economic, social/societal and ecological impacts of your organization that significantly influence the assessments and decisions of your stakeholders.
- **Completeness**
  Describe all material economic, social/societal and ecological impacts resulting from the activities of your organization that are considered important and that could potentially have an influence on the decisions of your stakeholders.
- **Balance**
  Provide a balanced and unbiased picture of the performance of your organization. This includes both positive and negative aspects (e.g. environmental pollution or noise generation).

- **Comparability**
  Compile your information in such a way that you and your stakeholders can analyze changes over time and compare them with other organizations.
- **Accuracy**
  Be as accurate and detailed as possible with your information and data, so that your stakeholders can also evaluate your performance.
- **Actuality**
  Ensure regular, annual reporting and the timeliness of the content. Do not repeat things that you have already reported without showing any change.
- **Clarity**
  Make sure that your information and statements are clear and understandable
- **Reliability**
  Prepare your report in such a way that it can be subjected to an external review. It is very important that your stakeholders are convinced that an external review of your report would confirm its accuracy (which does not mean that such an external review has to take place; that is entirely up to you).

9.3   Compilation of the Standard Disclosures According to GRI

First, you need to identify those of the general and specific standard disclosures that are required for our "in accordance—Core" option. Collect all the necessary data, documents and evidence in your organization and insert them into the structure and order defined by the GRI (see also Appendix 9 and 10).

Consider the principles of sustainability reporting in your work. If you are not sure what this means in detail, just read sect. 9.2 again and note down the characteristics that apply to you. The identification of material aspects and elements is of particular importance, as they form a central element for the "In accordance" option.

In contrast to the general standard disclosures such as strategy, organization, stakeholder engagement, industry-specific disclosures, etc., the specific standard disclosures are about the organization's management approach (Disclosures of Management Approach, or DMA for short) and indicators in all three dimensions of sustainability (economic, social/societal and environmental). The environmental indicators on resource consumption (material, energy, water, etc.), as well as the topics of wastewater, waste and emissions (especially of greenhouse gases), are of particular importance. In Chap. 4 we looked at our consumption behavior and already found there how great and increasing importance these indicators have for our purchasing behavior. It is therefore all the more important for every organization to report transparently and openly here and to be able to demonstrate significant improvements over the years.

Defining and continuously adjusting the material aspects and boundaries is a process that, if properly installed within the organization, enables a high-quality sustainability report. The cyclical process consists of the steps of identification, prioritization, validation and the final review. The results then control and regulate the first step again, see process in Fig. 9.1:

Following the individual process steps is essential for implementing the reporting principles, which are of central importance for achieving the desired transparency of the sustainability report.

Fig. 9.1 Reporting process

Step 1: Identification of Sustainability Context

First, the relevant topics for the organization are identified. If your organization has no experience with sustainability reporting, you should first focus on two topic areas:

- The internal structures of the organization
- The stakeholders of the organization

What are the relevant points in both topic areas? These are things that have a significant impact on the economic, social/societal and environmental elements of your organization. In addition to the internal structures, these include the detailed topics that have a great influence on the decisions of your stakeholders regarding your organization.

To identify these points, you should ask yourself the following questions:

- How are you seen and judged by your customers?
- How does a typical customer of yours behave when you increase or decrease the prices for your products or services?

- What happens when you bring new and innovative products or services to the market?
- How do your customers react when you switch to renewable energy in production or emphasize sustainability in the supply chain?
- How easy is it for you to recruit staff? Is your organization attractive for young professionals?
- How do incentive models for employees affect motivation and sickness rates?

> **Tip**
>
> Do a small workshop on this, invite as many creative colleagues from all areas of your organization as possible and do a little brainstorming. You will be surprised by the number and diversity of the different topics.

Step 2: Prioritization of Materiality

The GRI definition of the materiality principle states:
"The report should cover aspects that:

- *reflect the essential economic, ecological and social impacts of the organisation or;*
- *significantly influence the assessment and decision of the stakeholders"*

You have to analyse and prioritise your topics as far as possible. Continue your workshop and discuss the respective assessments with your colleagues.

If you cannot quantify a topic, that does not mean that it is not important and relevant. Take, for example, the topic of customer satisfaction. You have no evaluation of it? Nevertheless, you discover that this topic is obviously very essential for the development of your organisation? A fruitful strategic process begins.

- How can we measure customer satisfaction?
- What are the bottlenecks here and what are the main expectations of our customers?

You see, just dealing with the essential contents of your sustainability report triggers your complete sustainability management!

At this point, you can also specifically ask your stakeholders which contents and topics are important for them in relation to your organisation and should be included in the report. If you get feedback at this point, then I recommend you to include these topics in your report.

In addition, there are also stakeholders in your organisation that you do not ask, but should consider out of your social responsibility. First of all, there is the future generation:

- Does the activity of your organisation impair the resources of the next and the following generation?
- How do you affect the environment, the ecosystems and climate change?

When you have identified all the topics and aspects that are relevant and material for your sustainability report (as a result of your by now probably one-day workshop), then rank them in order of priority. To do this, evaluate the opportunities and risks arising from the topics according to the following criteria:

- the probability of occurrence of an impact
- the severity of an impact
- the probability of occurrence of opportunities and risks in connection with an aspect

- the significance of the impact for the long-term economic performance of the organisation
- the opportunity for the organisation to grow or gain an advantage, perhaps even a unique selling point, from the impact.

A very effective and yet simple tool for determining the order is a visualisation similar to the SWOT analysis already discussed (see Appendix 1).

Draw this simple quadrant diagram with the labels of the axes as in Fig. 9.2 on a flipchart or a sheet of paper and write the aspect, the topic above it (e.g. attractiveness of the organisation). Now give each participant of your workshop a sticky dot for each topic or aspect and ask them to stick their dots at the point in the diagram that corresponds to their assessment. (Use a different colour for each topic and make a small assignment list.) The division into four quadrants is not important yet.

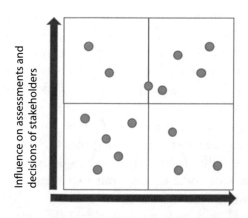

Importance of the economic, environmental and societal/social impacts of the organisation

Fig. 9.2 Importance and value of sustainability topics

The more significant the aspect is assessed in terms of its impact on the organization, the further to the right the point must be glued. The greater the influence of the topic on our stakeholders, the higher the point must be glued.

> **Tip**
>
> If you want to evaluate a sensitive topic in this way, for example a personnel issue such as gender equality in your organization, then simply turn the flipchart to the wall, glue the first point yourself and then let the colleagues glue their point individually.

Next, we can start with the evaluation; now the quadrants become important. We can safely forget the points in the lower left quadrant. The corresponding topics have neither a great impact on our organization, nor are they relevant for our stakeholders. Away with them!

In contrast, all topics and aspects in the upper right quadrant are of the highest relevance for us. These topics are all set for our report! We can also see an order within the quadrant. For the remaining quadrants, we have to take a closer look. Discuss with your colleagues whether the points should be included in the sustainability report, as they either have a significant impact on our organization or a great influence on the decisions and assessments of our stakeholders.

At this point, the top management of your organization is definitely required. The management (or the board) must fundamentally decide on the inclusion of the topics list you have developed. Ideally, your management participates in your workshop and can decide on the spot. At the end of this work step, you will definitely have a complete and prioritized list of topics in the order that applies to your organization (see Fig. 9.3).

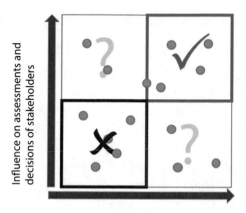

Importance of the economic, environmental and societal/social impacts of the organisation

Fig. 9.3 Weighted sustainability topics

If you have too many topics and aspects that you need to prioritize, then I recommend that you form clusters. There are of course countless ways to cluster sensibly, and you may already have the best division for you in mind. If not, then I suggest using the three dimensions of sustainability here. Form groups for economic, ecological and social/societal topics and sort the individual points according to their main focus.

Step 3: Validation of Completeness

We are now not far from creating and publishing our sustainability report and only need to check the collected information for completeness and at the same time ensure the involvement of the stakeholders. At this point, you should consider whether you should not have these checks carried out by an external consultant. This costs money, yes, of course, but it also brings you undeniable advantages. Of course, it also brings advantages to the external consultant, such as me, namely orders; but that's okay, I think.

- An external observer has a completely different perspective than you. Unlike you, he is not "blind to the operation", which you can never completely avoid, no matter how hard you try to be objective. An external consultant "sees" things that you simply cannot see because you are inside the organization. These new aspects and topics that you gain through an external observer can be very valuable for your organization.
- A good external sustainability manager has specialized in the topic of sustainability and is (hopefully) an expert in this field. Especially if you are at the beginning of sustainability and maybe designing your first sustainability report, you can learn a lot from a professional.
- You show your stakeholders that you are striving for objectivity and transparency and that you do not want to engage in greenwashing with your sustainability report.

Step 4: Publication

As a result of the steps identification, prioritization and validation, you are now ready to create and publish your sustainability report. You have meticulously and systematically collected all the data, topics and aspects and now have to put them into an easily readable and printable version.

Even though layout and design are not unimportant, please consider this argument: If you have this work done by an external specialist, and maybe even in a correspondingly high-quality execution, then this can easily cost a medium-sized company 50,000 to 70,000 EUR. Believe me, the content is more important for your stakeholders—in most cases you can create your first sustainability report in-house in terms of content.

Distribute your sustainability report to all stakeholder groups of your company (and of course also to your staff) and do not forget that you will be measured by it in the following years!

Post your sustainability report as a PDF file for download on your homepage and do not forget your reporting obligation to the GRI, if you have created your report "in accordance" with their standards.

Step 5: Feedback Round, Checking for New or Changing Content

Let us be clear about what needs to happen after the publication of your report, in order to anchor the topic of sustainability reporting sustainably in your organisation.

The day of publication of your report is day one of the preparation of the report for the next reporting year and you will most likely need this time to review your current report and optimise it for the next time.

Remember our virtual workshop on topic identification and prioritisation? You have certainly encountered new topics, new aspects and new relevances and materialities in the real implementation of this step. You cannot and do not want to simply leave these things as they are. On the contrary, you incorporate these new insights into your organisation and its processes and thus serve a key process of your sustainability management. You optimise the organisation with your colleagues and in coordination with your management! Do not forget to also obtain the reactions and statements of your stakeholders and to give their comments the same high attention as your internal reactions and actions.

Another review step concerns the specific characteristics of your industry. Also check whether there are aspects and specific standard disclosures for your industry. If so, apply them in your sustainability report. You can find these industry disclosures under Sector Guidance[6] on the GRI homepage. Currently, you will find information (Sector Supplements) for the industries:

- Airport Operators
- Construction and Real Estate
- Electric Utilities
- Event Organizers
- Financial Services
- Food Processing
- Media
- Mining and Metals
- NGO
- Oil and Gas

I myself participated in GRI-Supplement workshops for the Airport Operators sector in 2009 and 2010. As a reward for my work in this multi-stakeholder working group, which I enjoyed very much and where I learned a lot about sustainability, I am immortalised on the corresponding page of the GRI homepage under "Who developed the Supplement?"[7]. You can download the result of our work at https://www.globalreporting.org/search/?query=airport#:~:text=Airport%20Operators%20Sector-,Disclosures,-Reporting%20Resources if you happen to work for an airport operator or just want to see how industry-specific standard disclosures are defined.

I think you will find yourself among the ten industries and be able to extract specific information for your report from there.

When compiling your data and information, always pay attention to the materiality and do not inflate your report unnecessarily. Aspects that are not considered material for your organisation and your industry should not be covered in your report. In addition, you can and should of course include information on topics that are not listed in the GRI guidelines, but that are important for your organisation in your opinion, in your sustainability report. After all the general and specific work steps, we are now ready to create and continuously optimise our sustainability report annually.

9.4   GRI Standards

The GRI guidelines are constantly being further developed and so there is now a fundamental system change from the previous G4 guidelines to the new GRI standards.

The previous GRI guidelines, especially G4, were not flexible enough to form simple and suitable interfaces with other frameworks, such as the international Sustainable Development Goals (SDGs).

There are some significant changes in the GRI standards compared to the previous guidelines. The impacts of the reporting organisation's business activities (impact to environment), the aspect of materiality and the management approaches are more clearly taken into account or specified. The contents remain largely the same as in GRI G4, but are structured differently.

The contents of the GRI standards are modular and are divided into 37 modules. There are universal standards as well as topic-specific standards in the three dimensions of sustainability: economy, ecology and social.

The universal standards GRI 101—Foundation, 102—General Disclosures and 103—Management Approach are now mandatory requirements in the GRI Standards, i.e.

a sustainability report "in accordance" or "in core accordance" must fully include the Universal Standards.

The topic-specific standards GRI 200—Economic, 300—Environmental and 400 Social apply as far as necessary to understand the relevant sustainability topics in the organization.

Reporting Principles of the GRI Standards

Compliance with the reporting principles is fundamentally important to produce a high-quality sustainability report. According to GRI Standards, there are the following ten reporting principles for the content and quality of a GRI-compliant sustainability report (see Fig. 9.4):

Materiality in GRI Standards

Materiality has a special meaning in the GRI Standards. The sustainability report should therefore include in particular topics that:

- contain the significant economic, environmental and social

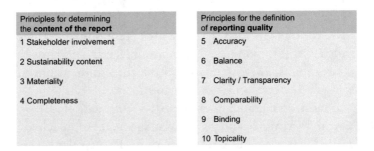

Principles for determining the **content of the report**	Principles for the definition of **reporting quality**
1 Stakeholder involvement	5 Accuracy
2 Sustainability content	6 Balance
3 Materiality	7 Clarity / Transparency
4 Completeness	8 Comparability
	9 Binding
	10 Topicality

Fig. 9.4 Reporting principles

- impacts of the organization and also
- substantively include the assessments and decisions of the
- stakeholders

You already know the materiality matrix in Fig. 9.5 from the previous chapter on topics and aspects of sustainability reporting. Here we look at the graphic again from the perspective of materiality. All topics in the upper right quadrant are material for us and thus set for the sustainability report. The topics in the lower left quadrant are not important for us. For the remaining two quadrants, you now have to decide on a case-by-case basis whether the respective topic is material for your reporting. The materiality of the topics changes over time for each company. Therefore, you should also review and adjust the materiality matrix regularly if necessary.

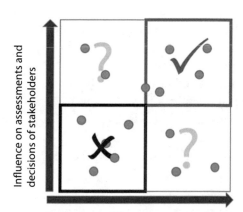

Fig. 9.5 Materiality matrix

Transition from GRI G4 to GRI Standards

An organization that still reports according to GRI G4 should not have major problems with the transition if the new structure and the partly new logic in the organizational areas are adopted and applied.

It should be noted that since July 1, 2018, the new GRI standards must be used in sustainability reporting if the report is prepared and labeled as "in accordance" or "core in accordance" with the GRI.

In summary, the following differences should be considered when switching to GRI standards:

- A report according to the GRI standards as a standalone sustainability report must contain a GRI content index.
- There is now a clear distinction of information and disclosures according to mandatory requirements, recommendation and guidance.
- In GRI 101 (Foundation), the new perspective of impacts (effects) on economy, environment and society is explained.
- In GRI 103 (Management Approach), the new focus on materiality becomes clear. The company has to explain the materiality of the organization and indicate the respective system boundaries

After that, you should perform the following transformation steps:

- Conducting a renewed materiality analysis based on the new priorities in GRI standards

- Checking which indicators, contents, data and KPIs can be transferred unchanged from the GRI-G4 world and where modifications are necessary
- Description/definition of modified/new indicators and KPIs
- Transfer of the previous G4 contents to GRI standards
- Checking the completeness according to GRI standards

For the transformation of existing G4 reporting structures to the new Sustainability Reporting Standard (GRI-SRS), the Global Reporting Initiative provides a very useful Excel tool[8] free of charge on its Resource Download Center in the area of "GRI Standard Resources". This should make the transition in the reporting system manageable for you (see also Appendix 10).

9.5 Non-Financial Reporting—CSR Directive

The law[9] to strengthen the non-financial reporting of companies in their management and group management reports (CSR Directive Implementation Act) to implement the EU Directive 2014/95/EU has been in force since the beginning of 2017.

The directive covers large companies of public interest that are capital market-oriented, have a balance sheet total of 20 million EUR, sales revenues of 40 million EUR and 500 employees or more. Credit institutions or insurance companies are subject to reporting if they employ more than 500 workers on average per year.

A company that is in principle subject to reporting is exempt from the obligation to prepare a nfB if:

- the company is included in the group management report of a parent company and
- this group management report is prepared in accordance with Directive 2013/34/EU

The company subject to reporting can use national, European or international frameworks for the preparation of the non-financial reporting → e.g. GRI standards.

The company subject to reporting does not have to provide information on future developments or matters if these are likely to cause significant disadvantage to the company and the omission of the information does not prevent an understanding of the company's business activity.

Requirements for the non-financial reporting

The non-financial reporting can be integrated in the annual report or published parallel to the annual report no later than four months after the balance sheet date. In case of an online publication, the non-financial reporting must be made available on the company's website for at least ten years. There is no audit obligation.

The non-financial reporting must briefly describe the business model of the reporting company. It also covers at least the following aspects:

- Environmental matters (GHG emissions, water consumption, air pollution, renewable energies, …)
- Employee matters (equality, working conditions, unions, health, …)

- Social matters (local and regional dialogue, protection and development of local communities)
- Respect for human rights
- Combating corruption and bribery

Information that is necessary for understanding the company must also be provided:

- The company's concepts, including due diligence processes[10]
- The results of the concepts
- Significant risks arising from the company's own business activities and how they are handled
- Significant risks arising from the business relationships and how they are handled
- Significant non-financial performance indicators for the business activity
- References to amounts reported in the financial statements, as far as necessary for understanding

Materiality According to the CSR-Directive/non-financial reporting

The materiality (see Fig. 9.6) of the non-financial reporting consists of five aspects and two risks of the reporting company:

Aspects
1. Environmental concerns
2. Employee concerns
3. Social concerns
4. Respect for human rights
5. Combating corruption and bribery

Materiality according to GRI standards	Materiality according to non-financial reporting
Material topics provide a balanced picture of the company's **material** issues, the related **impacts** and how the company deals with these impacts.	**Aspects** of the company's environmental, employee and social concerns, respect for human rights and the fight against corruption and bribery
Impacts the effects of the company on the economy, the environment and society	**Risks** that are very likely to have or will have serious negative effects on the **aspects**

Fig. 9.6 Comparison of the materialities according to GRI and non financial reporting

Risks

1. Significant risks that are associated with the company's own business activities and that are very likely to have serious negative impacts on the aspects, as well as the company's handling of these risks.
2. Significant risks that are associated with the company's business relationships, its products and services and that are very likely to have serious negative impacts on the aspects, as far as the information is relevant and the reporting on these risks is proportionate, as well as the company's handling of these risks.

Even if (so far) only large companies in the European Union are obliged to prepare a non-financial reporting, every company should actually recognize the opportunities that arise from the non-financial reporting. The image of and on the own company is sharpened and the trust of the stakeholders in the company is strengthened. In the context of a sustainability reporting according to the new GRI standards, a complete and precise tool for managing sustainability is created.

Compulsory or freestyle?

Whether sustainability reporting is an obligation for companies, or based on voluntariness, is not really important in my opinion. The decisive question is how we want to use the enormous possibilities that a sustainability report brings us internally and externally. In my experience, a sustainability report is an opportunity that no company should miss.

Now that we know how important the preparation of a sustainability report is and how it works in principle, the last section of this book begins.

We have acquired the necessary knowledge about strategies, goals and methods around the topic of sustainability. The preparation of a life cycle assessment and a CO_2 footprint, the importance of indicators, the sustainability KPI, we have learned as important elements for communication and control. With the GRI standards and the non-financial reporting, we can now present, report and further develop the achievements of our organization in a profitable way for us and our stakeholders.

However, in order to be able to implement all this knowledge in reality, we need a special type of person. A person who has the skills, the enthusiasm and the will to make sustainable structures a reality. We need the sustainability manager.

Source Reference and Notes

1. https://www.globalreporting.org/search/?query=reporting #:~:text=with%20the%20GRI-,Standards,-pdf
2. database.globalreporting.org
3. https://www.globalreporting.org

4. https://www.globalreporting.org/standards/resource-download-center
5. https://www.bmjv.de/SharedDocs/Gesetzgebungsverfahren/DE/CSR-Richtlinie-Umsetzungsgesetz.html
6. careful examination and analysis of a company, especially with regard to its economic, legal, tax and financial situation

10

Make your Dream Come True: Sustainability Manager

How do you become a sustainability manager, sometimes also called a CSR manager[1]?

In my opinion, there is not one way to become a sustainability manager, the tasks and requirements are too diverse. First of all, the general characteristics of a manager such as responsibility, decisiveness and negotiation skills are required. As a sustainability manager, you also need the desire to do something for society. Social responsibility, that is what the whole ISO 26000 is about, do you remember?

As a sustainability manager, you need to be highly innovative, both on the technical side and in process management. I would not be surprised if you, like me, came to the topic of sustainability as a lateral entrant. There are now many training opportunities and entire courses of

---

[1] CSR: Corporate Social Responsibility.

© The Author(s), under exclusive license to Springer-Verlag GmbH, DE, part of Springer Nature 2023
M. Wühle, *Making Sustainability Measurable*,
https://doi.org/10.1007/978-3-662-66715-6_10

study around the topic of sustainability/CSR, but most of the successful managers in the field of sustainability that I know are lateral entrants.

Why? Because a successful sustainability manager needs a lot of professional experience in as many different positions as possible and a stable value system. This mix of knowledge makes it much easier for him to engage with a multidimensional construct like sustainability. I do not mean to say that it would not be possible to become a successful sustainability manager after a successful study or training in the field, but it is simply harder without years of practical experience.

A sustainability manager also needs a high frustration tolerance, because it is especially difficult in this area to break up entrenched structures and achieve results. And even more difficult is to maintain what has been achieved against the recurring discussion about the meaningfulness of sustainability measures, especially when boards change and a new leader comes into the organization.

> But let this be said to you: Be cheerful and not needy of the services that come from outside, nor needy of the peace that others grant. You must stand upright, without being held upright. Marcus Aurelius, Meditations

10.1 Living Values

What do values mean for a sustainability manager?

Values are the guide rails that we anchor our activities to. Values reflect our basic conviction about important aspects in our private and professional lives. In relation to companies, the term Purpose is often used here (see chap. 6.1). Values are not creeds, they are also not goals. Values are the foundation on which we sustainability managers stand and which enable us to successfully implement our sustainability projects against all the resistance that we encounter.

Values are not goals. Our values help us to 'stay on the ball', even if we do not achieve our goals and intermediate goals, or simply cannot imagine achieving them. Values are our basic conviction about the important things in our lives. Everyone has their own value system. It exists regardless of whether we are aware of it or not. At this point, however, you should take the time and consciously bring your own value system to mind.

Let's do a little exercise again: Think for a few minutes about your values in your private and professional life. Then take a sheet of paper and write down your most important values (5–10). Values, not goals. A goal would be, for example: I want to become a millionaire within a year. The value behind it would be: My financial independence is very important to me. This allows me to realize my wishes. So go ahead, sit down at your desk and start documenting your values!

Are you done? If you have never done this exercise before, you are probably surprised by the result. Just like I was surprised. The result of my most important values looked like this:

- be aware of my experience of the here and now, be open and curious about new experiences.
- be open and accepting towards myself and others, towards life with its victories and defeats, and towards failure.
- be fearless; remain persistent in the face of fear, threat or difficulty.
- be kind to myself, take care of my health and well-being of physical and mental nature

- be aware of my own thoughts, feelings—especially my fears—and actions.
- take care of myself, choose my own way of doing things and persistently continue despite problems and difficulties.
- continuously work on improving, strengthening and promoting my skills and abilities.

Knowing my most important values has certainly stabilized my work and that will also be the case for you. Now you might say that values are important for every human being and not only for sustainability managers. I completely agree with you. However, the awareness of our own values helps us sustainability managers in a very special way. They give us the strength to keep going, even when, as is often the case with sustainability projects, the difficulties and problems increase enormously during the course of the project.

Values are something completely different from creeds, which only cement a system of wishes. Nevertheless, we have to deal with the creeds of our fellow human beings.

10.2 The Power of Creeds

We have already discussed in sect. 6.7 the peculiarities of municipal sustainability management and the often very obstructive creeds there. Let us put ourselves again in a typical situation for the postulation of creeds.

Imagine you are talking in the context of a project meeting about the next steps, your plans and ideas for the success of the project. Then someone speaks up and says that he does not believe that your concept works.

"Why", you ask, "do you not believe that it works?" The answer is usually something along the lines of "…

I've heard it from so many sides that it does not work …", "… my brother-in-law is a technician, he says, that's all nonsense …", "… everyone actually says that it does not work, and that's why I do not believe it either …", "… if at all, then I believe it only when I see it in writing …", or worst of all "… I just do not believe it, it just can not work …".

Does that sound familiar to you? Yes, I thought so. From my experience of many projects, you encounter such a person, usually even several, in almost every project. On the contrary, in a quarter of a century of project management, I can not remember a project where this phenomenon did not occur, even with intensive thinking.

What do we as project managers usually do in such a situation? If we are inexperienced, we simply ignore these skeptics. We refer to our mandate and refer the annoying troublemakers to the next higher authority. We have already addressed this situation in the chapter on sustainability management in companies.

It is quite convenient, sometimes even successful, to involve the board or the management, but not optimal. Because if this skeptic can not (directly) nag by order, this person will work against the project in the underground with 99.9% probability and hinder you in the easiest case. In the worst case, this person will defame and hurt you as a person and human being by rumors, mobbing, delaying tactics, latrine slogans and even uglier.

If we are already experienced managers and have already paid a lesson in this respect, then we will try it differently. We then try to convince this skeptic with logical arguments and evidence of the correctness of our concept, and use a large part of our energy (and time) for a persuasion work that is completely pointless.

It is pointless because we are dealing with creeds in these cases. If a person firmly believes something, maybe even fanatically believes it, then you can not break this belief with logic.

There is the relatively low chance that you convince such a 'believing' person by the practical application. Sometimes even a surprising change of faith occurs and the person concerned says afterwards that he actually always believed in the success.

More likely, however, this person will stick to their original belief and explain your practical success with exceptions that made this and only this one success possible.

If we want to be successful as sustainability managers and also have fun and joy in our work, then we need a special knowledge and methods to deal with our fellow human beings, whom we often encounter as skeptics in the project, and who confront us with their creeds.

As a project manager, you always face these creeds and have to cope with them. As a sustainability manager, this phenomenon will occur even more, because as a sustainability manager you are always on the road in a very complex network. You do not just manage a simple project (which can be complicated enough), but move in a multi-dimensional field with numerous interrelationships.

That you step on the toes of different people is almost inevitable. You can not escape that. The question is, how can you best behave towards fixed creeds?

Ignoring and delegating upwards usually does not work in the long term, we have understood that. We can not convince 'believers' either, their 'faith' is much too strong. How do we deal with the creeds of our skeptics? How should the roles be distributed? What other tools do we need for our tool backpack and where do we reach real limits? We have to clarify these questions before we are fully equipped as sustainability managers to be able to work successfully.

10.3 The Role of the Sustainability Manager

First of all, we have to keep in mind our role as sustainability managers and constantly have it present at the surface of our consciousness.

We are aware of the need to spread the concept of sustainability as widely as possible, especially in the age of global climate change. However, we are not missionaries who want to impose their creed on humanity. We offer solutions for very complex problems. As mediators between technology, economic requirements and the people involved, we find the right approaches from case to case. However, we are not preachers who want to convince other people of their solutions.

We are aware that in the complex environment in which we operate, our customers/clients are the pace-setters for change and innovation and not us. We advise and help people who have already developed a clear feeling on their own that their organization needs to be changed if it is to have a positive development.

Our clients must also have developed their own ideas and goals and fully stand behind them. We can help these people to a large extent with our knowledge, our experience and our special methods to achieve their goals. We round off the edges and corners of the self-developed ideas and goals and work out a sustainable structure for their organization with our clients.

Persuading others is not our task, but that of our clients. If necessary, we give them the necessary things that are needed for their inner persuasion. Our position is in the second row, because we want our clients to identify with the necessary changes and innovations and to further develop them when we are no longer at their side.

We have to internalize this role understanding and also communicate it clearly to our customers/clients from the beginning. Our customers are people and not robots. Therefore, we have to think about some human aspects and about special methods, how we have to deal with people especially in the field of sustainability.

10.4 Dealing with Human Behavior

We have to be clear that the people we deal with in our projects are also our greatest resource. If these people come to the conviction on their own through our work, our appearance and our concept that sustainability is the right way for their organization and they identify with it, then the race is already half won.

Conversely, the same applies. For us sustainability managers, this means talking to as many stakeholders as possible, involving them closely in the project from the start and, above all, listening to them attentively. Listen carefully and remember who mainly tells you about problems and difficulties and who has already thought about possible solutions. The people of the first group still need time. We cannot and do not want to convince them of something that goes against their creed. Some people of this group will change their creed on their own in the further process and then they are ready for us.

The people of the second group are the ones we have to look for and find especially at the beginning. This second group is numerically much smaller than the first. However, they exist in every organization and sometimes you just have to lure them out. This can be done, for example, by having the management of the organization (board, managing director, mayor, etc.) form a special project team that reports directly to the management and also receives its assignments from there. This leads to a status gain for the members of the project team, which in turn attracts more suitable project members. If you also make it clear from the start that this project team will work in a committed and solution-oriented way, then usually only the people of the second group will join the team.

It is also important that we do not have to understand or know everything in order to apply it. This applies to you as a sustainability manager as well as to the people in your project team and the management of your organization. Very many people can drive a car, a very complex machine. But most of these people do not know how a combustion engine or an electric drive really works, or how to tell if the battery is still okay. This is not necessary at all. What is important is that we can operate the car and that we know which mechanic, which workshop we have to go to if there are problems with our car.

This applies to the same extent to us as sustainability managers. We do not and cannot know all the technical details in the affected organization, let alone be competent there. We only have to be able to steer the process, the car and know who to go to for which detail.

Fear of the new

A sustainability manager always brings something new to the organization he is working for. This is in the nature of things, because until today almost all organizations are strictly set up according to business management rules. This has worked well in the past and now we come and claim that organizations have to set themselves up according to sustainability criteria in the future if they want to assert themselves and survive. At this point we need a lot of understanding for the human behavior that comes now.

For the majority, new things mean a threat to the daily routine. New things make them afraid. Fear of change in general, fear for their jobs, fear for their status, fear wherever they look. We sustainability managers have to take these fears very seriously if we want to be successful. Even and especially when these fears seem completely absurd to us, we have to deal with them and try to reduce them. If we don't, then these people will work against us with all their strength, with all their intelligence and most likely block our work completely in the end.

Always remember, people are our greatest resource. They have to be at the center of our work, even and especially when we would rather focus on technical solutions and exciting innovations.

Why not, technology is simple and fun. If you have a technical profession like me, then you know this from your own experience. Implementing the found technical solution economically is considerably more difficult. But experienced managers like us also manage to do that sooner or later. Getting the affected people to accept and adopt the found solution is by far the biggest problem.

I once had the task as part of a very professional project team to optimize the fleet of a company and make it fit for the future. The management was behind us and that would make resistance actually pointless, one would think. After a thorough analysis, we developed a fleet management system that was easy to use and economically highly efficient. But when we handed over the system to the fleet managers, it was clear to me in an instant that our work of the last months would remain completely unsuccessful.

I still remember a rather bizarre situation: In a meeting room, the fleet managers and we from the project team sat opposite each other. The body language of our counterparts signaled maximum resistance—crossed arms, the body turned sideways, avoided eye contact, at most a forced smile that could not really hide the dislike and the defense.

Have you ever experienced something like this? Yes, then you can also imagine what came next. The fleet managers accepted the work results and the system documentation, because they could not refuse that due to the decision of the management. They thanked us laboriously for the work done and left the meeting room after a fleeting handshake. Less than two months later, when the project was finished and our project team no longer had a mandate, they threw the new system in the trash and returned to the old and inefficient system.

That annoyed me terribly at the time. All the work was for nothing! Today I understand the mechanisms that led to this refusal attitude. We relied on our mandate and ignored or underestimated the worries and fears of the affected people.

And that's what I want to give you at this point. Have understanding for human behavior! Put yourself in the position of the affected people. Go back mentally to a time when you may have experienced something similar. Try to imagine a situation where something new is imposed on you and you don't want it and are also afraid of it.

Ask yourself: What do I have to do in this situation to get the participants to cooperate and accept the new thing?

10.5 NLP and the Power of Questions

A very excellent method to overcome the power of the creeds, to better understand the people around you and to be able to respond to their worries, fears and wishes, offers the neuro-linguistic programming (NLP). It would certainly go beyond the scope of this book and also my own knowledge to want to introduce this topic more comprehensively here. Nevertheless, a few sentences about NLP are allowed at this point.

I myself had contact with NLP for the first time a few years ago, when I was professionally in an extremely difficult and personally threatening situation. I got the recommendation at that time to take advantage of the help of an NLP coach, which I did immediately, because it was the straw I was desperately looking for. This NLP coach helped me a lot in this phase, especially by asking me the right questions, which then finally led to the solution of my almost existential problem.

I can definitely recommend such a coaching. In addition and complementary, there is a lot of good literature on neuro-linguistic programming, to which I would like to refer you. I read a lot on the topic at the time and acquired further valuable knowledge in self-study, which I would like to pass on here in brief, which can help you as a sustainability manager in dealing with your fellow human beings.

Before I start, however, I would like to recommend two books by Anthony Robbins, which were the actual key to success for me: "Unlimited Power" and "Awaken the Giant Within". I don't know Anthony personally and I don't get any (unfortunately) bonuses for recommending his works. But since his insights and his interpretation of the NLP methods were so valuable to me, I would like to make it easier for you to find a suitable access to the topic of NLP with these books.

But first something about the term neuro-linguistic programming, as I understood it. NLP is a scientific method with which we can change our behavior and the behavior of other people, i.e. reprogram them. It was developed by the US scientists Richard Bandler and John Grinder with the help of the most modern findings about the function and operation of the human brain. The two scientists analyzed the behavior of many successful people and developed the NLP principle from it.[1]

This reprogramming, which we can do with ourselves and partly also with other people, has nothing to do with magic or even with faith. That is the amazing thing about it. You don't have to believe in it for it to work and you don't have to develop a creed either. When you reshape the neural roads of your learned behavior, your brain will automatically adopt the new behavior. It works for sure and I recommend that you get involved.

Let us now start with the first NLP topic that I want to address here. It is the questions.

The power of questions
Questions have an incredible power and effect on our brain. When you consciously ask yourself a question, your brain cannot help but produce answers. Answers that are also subjectively true, because your brain cannot lie to you. These are basic functions that we cannot and do not want to turn off, but rather use them for ourselves.

Take some time now, sit down at your favorite place and ask yourself questions about your most pressing professional problems according to the pattern:

- What do I have to do to make my project successful?
- Which tools from my tool backpack can I use?

- What support can I get from outside?
- How can I get the people involved to support me?
- What are the biggest obstacles right now and how could I overcome them?
- What words do I have to use to be understood emotionally as well?
- What is the main reason why I have not been successful so far?

Write your questions on a piece of paper. On paper and not on the computer, because that is much more effective! Writing them down is already an NLP programming that works for sure. By writing, you visualize your questions, by writing with your hand you get an intense feeling and connection to your questions. Leave two or three lines free after each question. Then take the sheet of paper in your hand and read your questions again. Then, read your questions aloud to yourself again.

What kind of type are you?
And now take some time and think about what appealed to you most in your exercise.

Was it the fact that you wrote down your questions, transferred them from your brain to a piece of paper, and can now see them clearly visually?

Was it the feeling in your fingers that you had when you wrote down the questions with a pen, and the feeling of holding the sheet with your questions in your hand?

Or did the literal loud reading of your questions appeal to you?

Depending on that, you are either a visual, a kinesthetic or an auditory type. Mixtures are possible and common. I, for example, am clearly a visual type with kinesthetic facets. Seeing the questions on the paper and holding the sheet with the questions with my fingers is the most intense way for me to deal with my questions. And believe me, I do this exercise regularly with increasing success.

Have you already figured out what kind of type you are?

Now make a point of consciously using the form that you respond to most strongly in the future. And not only with our question exercise, but from now on in general with your work!

Back to our exercise. Read your questions again, quietly or loudly, depending on your type, and hold the sheet in your hand if you have kinesthetic components. Do you notice anything? Your brain produces answers to your questions. You can't prevent that!

Write these answers immediately under your questions. This is extremely important, because in such a one-person brainstorming your brain does produce answers, but they are often quite fleeting. You will not necessarily get an answer for every question that you can use the first time, but that does not matter at all. This is a process that you should repeat regularly, at least weekly, like me. Then your sheet of paper will fill up with answers and the space will soon not be enough.

Your questions will also change through this process, because the answers to your questions lead to further questions. I like to compare this process to the labyrinth of the Minotaur (see Fig. 10.1). Each answer brings us closer to the exit, the solution of our problems.

NLP = sustainable reprogramming

I assure you, if you do this exercise regularly, you will reprogram your brain in such a way that after a certain time it will spit out answers again and again, even if you are not consciously thinking about your problem questions.

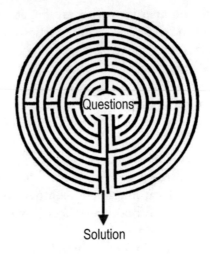

Solution

Fig. 10.1 Questions lead to the solution

> **Tip**
>
> Enhance the effect by memorizing your questions for a few minutes before getting up in the morning. In this relaxed state, not asleep anymore and yet not fully awake, your brain will surely produce one or two flashes of insight. Try it out, it works amazingly reliably!

With questions, we thus apply an elementary NLP program that runs completely independently of our consciousness over time. It is then firmly anchored in the autostart group of our brain.

With questions, however, we can also influence our fellow human beings. Do you remember what we said about the many skeptics and holders of creeds? We cannot and we do not want to convince them of anything, let alone the sense of sustainability. I used to make this mistake often and always failed. So we should avoid that.

However, we can ask these people questions, following the pattern:

"What do you think, how could we solve this problem?"

"What would you say is the better solution?"

"What do you believe is the reason why other organizations deal with the topic of sustainability?"

"What do you think, how can we reduce our energy costs?"

"Do you have an idea how we can motivate our staff more in connection with environmental and resource protection?"

"Do you see alternatives to the previous project approach that lead us to the same goal?"

Do you have a clue what happens through these and similar questions?

Exactly, the brains of these skeptics cannot help but produce answers. That's the trick.

If these people have themselves well under control, they may not give you the answers right away, but the program that you started with these questions cannot be easily stopped either. Not if you keep asking, then the process runs automatically similar to yourself. This way you increase the probability that more people will work constructively with you and your project considerably.

Reframing

Another very powerful NLP tool is the so-called reframing, the reinterpretation of terms in our communication.

We all know the example of the half full or half empty glass. No matter which of the two terms I use, I am talking about exactly the same thing. However, they have completely different meanings for us. A half full glass is usually positive for us, a half empty glass negative. "Too bad, not much left." If we consciously pay attention to calling the glass generally half full in our communication, we give negative topics a different, a positive frame.

We can reinforce this reinterpretation of negatively connoted terms with the use of a transformative vocabulary, another mechanism in neuro-linguistic programming. With the transformative vocabulary, the reframed term is emotionally anchored in the brain. To stay with our half full glass, this could then become something like:

The glass is still half full, that's great!

At the beginning it may seem strange to you to use such intense words and maybe you also think it is exaggerated. That is exactly the point. These reframed, reinforced terms trigger in the brain of the receiver exactly the emotion that we want to evoke:

Positive and solution-oriented thinking.

That's when even the skeptics start to wobble.

Take a few minutes and think about which terms and words you normally use when you communicate about problems and progress in your professional everyday life. Give these words and terms a new, positive frame and reinforce it to effectively program the transformation of the term:

From "I am satisfied with the project progress" becomes: "What has already been achieved is absolutely top-notch".

Instead of "I am happy about our successes" use: "I am totally thrilled about the successes we have achieved together".

Do not say: "This problem is frustrating", but: "I find this challenge fascinating".

From "I am unsure whether we can achieve this goal" becomes: "This is not quite in the bag yet".

Instead of "I am worried that ..." becomes: "We still need to work on that", or: "I need to question that more closely".

When someone asks you how you are doing with your (sustainability) project, what do you answer? Usually something like "it's okay", "pretty good", "it's going", or "so-so", right?

Use instead much more intense answers to the question of how you are doing, such as "incredibly good", "everything is first-class", "great", "it's going forward/ upward", "it couldn't be better".

Try it. Use superlatives. Two effects happen. You program yourself for success and motivate yourself additionally. Your project participants will also be positively tuned and your skeptics will be baffled and keep quiet for a while.

> **Tip**
>
> Make sure that you use the word **but** as little as possible, preferably not at all. When you answer a statement from someone else: **"Yes, but"**, then you send the other: **"No, you are wrong"**. This signal arrives irrevocably like that with us and causes defensive behavior. Use rather connections like **and**. From **"Yes, but there is still ..."** becomes then **"Yes and let us also think about ..."**

Reframing and transformative vocabulary are very powerful tools for our success as sustainability managers. One of my role models in this regard, mentioned already at the beginning of the book, Steve Jobs, has used this kind of communication in my opinion masterfully.

Take a look at some videos of Steve, in which he introduced his new products like iPad or iPhone. From an already great product he makes with his words something extraordinary and can present the special features of the devices very captivating and vividly.

Listen to the words he used:

incredible experience
unbelievable
truly remarkable
extraordinary
I was thrilled
pretty, nice (the design)
awesome
pretty amazing
far better at some key things

With this, Steve has reinforced the success of his products even more and we can do the same in our work as sustainability managers. I have been working according to this method for some years now and with increasing practice comes also increasing success. I notice this very clearly by the feedback that I get meanwhile on my lectures and presentations, with which I can often arouse real enthusiasm.

The conscious use of positive and constructive beliefs in the form of metaphors can be used to change the mental attitude. The change of one's own attitude and that of our fellow human beings, especially the dear skeptics. I use for example for myself and others very successfully:

There is no failure, only results.

I have formulated these and the following beliefs for myself and memorize them almost daily:

Beliefs that Lead to Success

- Everything happens for a specific reason and purpose and can be of benefit to me.
- There is no failure. There are only results.
- I take responsibility for whatever happens.
- I don't have to understand everything to use it.
- People are my greatest resource.
- Work is play.
- There is no lasting success without commitment.

Tip

Formulate your own beliefs. Not necessarily the ones you have been following for many years, but constructive beliefs that give a different, positive meaning to the events around you and move you forward on your path.

These beliefs give each event a positive quality that is otherwise often seen as negative and frustrating. It is a form of reframing and linguistic transformation that has become normal behavior for me (that does not mean that I am not also sometimes depressed when not everything works out as I imagine it).

For me, the main approach behind this is to constantly try out new methods to achieve the desired success, and to recognize and optimize my own behavior in the process. When you have received a result from an action, pay close attention to your behavior that led to this result.

Did your behavior bring you closer to your goal? Then use this behavior again.

Did your behavior move you further away from your goal? Then change your behavior at the next opportunity.

Not my goals change, but my behaviors and the methods I use to reach my goal. Every carefully analyzed result of my actions brings me closer to the goal.

Neurolinguistic programming is certainly a very powerful tool that can certainly be used not only by sustainability managers to become even more successful.

Nevertheless, NLP is not a panacea and I really know enough 'believing' skeptics and consulting-resistant people, for whom even the most sophisticated psychological methods and tricks do not work. However, there are a number of other tools that can support us; I would like to discuss some of them in the following.

10.6 Investing in Networking

Networking is an indispensable help and support. There is already a solution for almost every problem and the trick is to get to this solution quickly and without great costs. Networks are very helpful in this respect. In a network there is a constant give and take and whoever thinks they can only take there will be very quickly isolated. Especially at the beginning, networking means serving the giving side more, but then you also benefit from it.

It takes some time until you can really benefit from your network. I compare networking with a savings account. You have to pay in for a while and also have reached a certain size before a significant profit becomes visible (back then, in the time when there were still interest rates on the savings). I don't want to claim that it takes 10 years like it did for me until you can really benefit from the network. That can certainly go faster, but some time has to be invested.

Quality is in Demand

I know people who have over 10,000 contacts in a network. Have these people become millionaires because of that, or even halfway successful? I have never met or even heard of anyone who is really successful with this kind of networking. The people with the 10,000 contacts are not the magical influencers, but are the people from whom we constantly receive completely annoying advertising mails.

It is advisable to build up your own network consciously and purposefully. I only accept contact requests from people with whom I have a substantive interface. I proceed exactly the same way with my own contact requests. It then takes a certain time, as I said, until real benefit can be drawn from the network, but it is worth it.

In order for networking to pay off in the end, it is necessary, besides the input that has to be brought, to also design your own profile on the network platform impressively. You can find enough good instructions on the internet on how to do this best. I may have also taken 10 years to really benefit from my network because I neglected this part in the first years. Therefore I recommend—even if time is always short—to invest enough time and good content in the design of the network profile.

> **Tip**
>
> With most networks you can start with a free access at first. If you like it, you can switch to the premium version. However, you should test extensively before you make this step, because the premium functions are not always worthwhile. Often you can also get a free month for the premium version.

10.7 Innovations are Promoted

Another tool is innovative techniques, methods and processes. Innovations are friends of every sustainability manager, because they bring fresh wind into entrenched routines and often allow quantum leaps in your project.

Where do you find these innovations? Very often through our deliberately built networks. In network groups—which we have chosen purposefully—information about innovative technologies is often discussed and presented before they are spread through the usual media channels.

Another good possibility for me is to visit trade fairs. With a little preparation, i.e. browsing through the trade fair catalogue and noting down exhibitors, whom I know have been innovative in the past, I make a rough route plan and then go. I also like to ask directly: "Do you have anything new, innovative?". Then often quite astonishing answers come, which often help me a lot in my concepts and plans.

> **Tip**
>
> If you run a blog, you can often get accredited as a press representative at many trade fairs. The advantage is—besides a free access to the fair—that you can get information about innovations from first hand quite easily in the press center of the fair.

Innovative technologies, especially in the context of digitization and energy transition, are financially supported to a large extent in some cases not only in the member states of the European Union. We know that if we can create incentives in the economic dimension, then our projects run much easier and skeptics even change their creed. Therefore we should use the existing funding programs whenever possible and sensible.

There are numerous funding programs for innovative technologies, especially in the fields of renewable energies and alternative fuels. So areas of interest in which we as sustainability managers often move. Often the access to these funding programs is easier than it seems. Don't let yourself be deterred by the (intentionally?) set up bureaucratic hurdles and submit the corresponding application. Sometimes it is also necessary to register and prove your expertise as an expert in the respective field. If the funding program fits your topic, then you should take this step. It will give your sustainability project the decisive impulse.

10.8 Project Management

A professional project management is of course always one of the most important tools for a successful project process. This topic is quite extensive, so I have to refer mainly to the numerous specialist literature. However, I would also like to suggest a few things about project management here.

Define the project goal together with your client as precisely and unambiguously as possible. Every project has a qualitative and/or quantifiable goal. Every project has a defined budget and every project has a temporal frame with a fixed start and end date. Describing the project goal in this way is also the first task for us sustainability managers, in which we should invest enough time and always write down the result.

Equally indispensable is a milestone planning, which we use both for planning the individual project phases and for result control. There are heaps of software, good and less good, you have to choose according to your own taste. There are expensive and free Project programs[2], which are all very extensive and powerful and with which you can control even the largest projects with all resources. However, I would also like to warn you at this point about the amount of work involved with such programs. In really large projects, the maintenance of these Project programs including personnel is provided. In most cases, in my opinion, you can also work with simple tables. In all the years that I have been working as a project manager, I have only worked twice with one of these Project programs and each time I had a colleague who did nothing but maintain these data and run evaluations.

Demand Decision-Making Competence

For you as a project manager, in addition to the project goal and the milestone planning, the consensual determination of your competencies and decision limits is indispensable. Many projects get into trouble or even fail because this issue is not clarified before the actual project start. Define for yourself the decision-making competencies that are indispensable for you and make it clear to your client that you can only take responsibility if you also have the possibilities to act responsibly. If they try to establish you only as technically responsible, but not as disciplinary and cost-wise, then you better leave it alone, if you can.

10.9 Outfit and Communication Style

Finally, a few words about our appearance and our personal impact. Clothes make the man, that is an old saying that still holds true today. A Steve Jobs could afford to celebrate his unique events only in jeans, white sneakers and a black sweater. This outfit matched his minimalist ideas and since he lived them, already had a high level of recognition and enormous professional success, he also came across as authentic to his customers with this clothing.

Clothes Make the Man

As we have already established at the beginning of the book, we are unfortunately not Steve Jobs and therefore have to follow certain rules if we want to be successful. This includes a well-groomed appearance, because we are unconsciously always judged, classified and stored by the people we deal with. You can see it as a NLP method, with which you can win over your fellow human beings. I do not mean that you always have to appear in a suit and tie or costume. This is sometimes necessary, especially for public appearances and appointments at board level. To come to such appointments in everyday clothes is not possible at all, because they will not take you seriously. If you are not sure what kind of clothing is welcome, then ask the organizer of the event/meeting openly about the dress code. What this means in detail you can find quickly on the internet.

We Talk With our Hands and Feet

When you then (appropriately dressed) start with your concern, pay attention to your language. For example, I tend to mumble, swallow words and also speak too quietly. **Focus on a clear and distinct pronunciation**. Speak consciously louder than you would normally do. This takes some overcoming at first, but it is a matter of practice. Make eye contact and underline your words with gestures. I do not mean hectic or nervous movements of the arms, but the underlining and reinforcing of your statements with calm and moderate gestures. Do not stay in one place, but move around the room. **The right body language convinces more than 1000 words**!

> **Tip**
>
> Practice your appearance. Watch videos of your role models and pay attention to their gestures and movements. Then make a video of your own presentation and compare your appearance with your role model. Change your style until you are satisfied.

Do not cross or fold your hands, because that looks either insecure, rejecting or arrogant. If this is too open, too insecure for you, especially at the beginning, then take something in your hand. For example, a presenter, a laser mouse or your manuscript cards. These objects are something like an anchor for the stage fright, on which you can hold on. Change this object from the left to the right hand and use it consciously to underline your arguments. You will see, a great tool, almost a baton for your audience!

We have to be aware that our appearance, no matter in what form, is registered and automatically evaluated by our fellow human beings. We should not let this unsettle us, but use the effect specifically for us (NLP).

10.10 Self-Realization Requires Discipline

After we have discussed the essential basics that are necessary for a successful sustainability manager, I would like to talk to you about the reward you will receive if you engage in an activity in the field of sustainability management. And we also have to say a few words about good and efficient work style.

Of course, as a sustainability manager, you will also want and earn money and, depending on your wishes, also achieve a certain level of prosperity. I will come back to that later. First of all, I am concerned with the ideal reward that you receive when you observe certain behaviors and consciously remind yourself of essential skills that you carry within you, time and again.

- **Serenity, diligence, dedication, concentration on one thing**
  It is not always easy to remain calm when our income depends on the success of our work, and I can very well understand that one becomes restless and nervous when success does not come immediately. Believe me, I can understand that very well, because I have experienced it myself and still experience it from time to time. Therefore, as much as possible, provide for reserves in times when you are doing well economically and invest these reserves profitably, but in such a way that you can have them available again within a few months without losses. These reserves help you to maintain the necessary serenity when things get difficult.
- **No pain, no gain**, you can underline and take that literally. It helped me a lot that I plan every working day in writing. At the end of each working day, I enter between two and five things in my calendar that I want to do the next day. No more and no less. In the meantime, I have so much practice and can assess my performance so well that I usually always manage. Every completed point gets a green check mark, which motivates me additionally. If I have not managed a point once, I cross it out and enter it again immediately on one of the next days.
  I have also learned that the talk about the multitasking ability of successful managers is complete nonsense. When we try to do different things at the same time, nothing clever comes out of it. I always focus on one topic, one task, and force myself to ignore everything

else until this one thing is done. Then comes the next point. With this method, I now work very efficiently and successfully.

> **Tip**
>
> Avoid parallel work. Focus entirely on the respective point and work consciously serially until the point is done. Only then follows the next task.

Every human being has phases during the day when work is particularly easy and concentration is highest. For me, this is the time between seven and twelve o'clock in the morning. The first hour after drinking coffee is almost a warm-up phase, in which I start to tick off the first points from my daily list. Nothing dramatic, rather simple things. Then, when I'm warmed up and my performance turbine is running smoothly at idle, I start with the most important point of the day. At first, I cheated a bit on myself, because I usually started a task that I enjoyed. In the meantime, I have learned that it is absolutely necessary to put the really most important task in this most productive time of the day, even if it is something that I don't like to do.

> **Tip**
>
> Find out your most productive time of the day. To do this, make a weekly review and note the times when you had worked most successfully and productively. Always put the most important things of the day in this time period. You reward yourself with a strengthened concentration ability and serenity that goes along with it!

- **Perseverance, joy of success, work is fun**

  Perseverance plus patience plus dedication definitely lead to the goal. Your perseverance will improve if you plan your workday as described and define concrete things that you want to work on. Be really specific about your work plan.

  Do not write: "Start creating concept xy", but better: "Define the structure for concept xy and provide it with headings". This way you have given yourself a clear work result that you want to achieve. When you get to this task, you can quickly deduce from this clear task description how you want to proceed, what steps are necessary for it and how much time you will probably need for it. When you have completed this task, you put the green check mark behind the point.

  It makes me happy every day anew when I see how I have worked through point by point, that is very important to me and the cornerstone of my professional success.

  Feel joy of success, because work should be fun. If you enjoy your work, you will also get the perseverance and dedication that are so important for success. Sometimes I feel a slight regret when I have completed a task that was successful and gave me a lot of fun.

- **Self-control, pride, self-confidence**

  In order to be able to persevere, to muster the necessary concentration, a high degree of self-control is required. In "Gorin No Sho", the book of five rings, the probably most famous samurai Miyamoto Musashi writes about four hundred years ago about the mindset that is required for success (of sword fighting):

  With an open mind and not rushed, look at things from a higher point of view. You must cultivate your wisdom and your spirit. Both in combat and in everyday life you should be determined, though calm. Meet the situation without tension and yet not negligent.[3]

Miyamoto Musashi stood in combat as if beside himself and observed action and reaction from a higher point of view. This reminds us very much of certain NLP methods, doesn't it?

I have translated this mindset for myself with self-control. It is not about being perfect. What would we be without our little weaknesses and quirks? The only important thing is to have a high degree of self-control at the times in everyday and professional life when we want to accomplish important things successfully and win the fight.

A higher degree of self-control can be learned by anyone who has the will to do so. Planning your daily work in writing and in concrete terms is certainly an important step towards this. You will notice over time that you become increasingly proud of yourself. The successful self-control creates pride and strengthens your self-confidence. You are the master of your actions and not the slave of your calendar!

When you look back and see what you have achieved, done, conceived and moved, you can rightly be proud. From this pride in the achieved high degree of self-control, an unshakable self-confidence emerges, which leads to further successes:

> What I set out to do, I achieve!

And thus the circle closes. With this great and justified self-confidence, you finally get the serenity, joy and dedication that are indispensable for a sustainably successful work. The accompanying positive emotions, the shiver that may run down our backs when we sense that we are approaching the goal, these emotions are the most lasting access to our personal resources.

As sustainability managers, we feel our social responsibility and that is joy and task at the same time. In the age of global climate change, we are challenged to play a decisive role in shaping the transition from an economic to a sustainable development. Through our work, we experience a high degree of satisfaction, because we know that we are doing the absolutely right thing, for ourselves and for the generations to come. Suitable for grandchildren. Besides, we must not forget that we also want to live from our work.

10.11 Making Money is not a Shame

Making money, not giving away the acquired knowledge, but marketing it!

With all the challenges and skills that a successful sustainability manager must have or acquire, we must not neglect the advantages that come with this profession (or calling?). Even though sustainability/CSR is new, incomprehensible or even misleading for many people, sustainability has a long tradition. Starting with Hans Carl von Carlowitz, organizations have been using methodical sustainability mechanisms for more than three centuries to achieve a lasting and socially acceptable success of their organization. Especially in the medium-sized sector, the "honorable merchant" still has a high value. This term contains much of what we understand today by a sustainability manager. The medium-sized sector is today receptive to managers who advise their company holistically and contribute to its good development.

Even large corporations and multinational organizations are increasingly paying attention to the social and ecological aspects of their suppliers and business partners. The interest of investors, analysts and rating agencies in sustainable structures in the economy is growing. I have witnessed myself how the creditworthiness of a corporation was set to triple-A (AAA) when it could be proven that the corporation had a sustainability strategy, regularly published a sustainability report and had corresponding sustainability goals in the target system of the executives.

This interest of medium-sized and large corporations is of course the great opportunity for us sustainability managers to get good fees. We are a sought-after and thinly sown group of experts and that is always the best prerequisite for good fees or remuneration. We are in demand because we can confidently cover a very wide range of social, economic and ecological facets. We can reduce the energy consumption of a company and at the same time implement ecologically sensible energy concepts. Creating a CO_2 balance and developing a CO_2 reduction path is a matter of course for us. Do you need to create a sustainability report? No problem, we show you how it works. Is a supply chain socially or ecologically questionable? We analyze the supply chain and give practical advice on optimization. Is it about establishing environmental and social standards in developing and emerging countries? This is an interesting task for us, because we are knowledgeable, experienced and innovative. We can show organizations and companies worldwide the benefits of their commitment on the economic, social and ecological sector.

We must not forget one thing and make one mistake, namely to ask for too little or no money for our services. In my experience, sustainability managers are extremely vulnerable here. They are all very socially minded people with a high sense of responsibility for society and the environment. Often we do something for free for idealistic reasons, for which we should actually charge ourselves.

I say this from my own experience and admonish myself. The knowledge, skills and methods that you have acquired in a years-long learning process and probably paid for with frustration and tears, must now also be used to secure your livelihood and fill your refrigerator.

I know a good sustainability manager with said high social commitment, who is satisfied with a daily fee of 500 € under the most difficult conditions in crisis areas, even though he knows that he is quite underpaid. There is no fee schedule for sustainability managers like the fee schedule for architects and engineers. However, I would like to draw a lower limit, which is based on my long experience as a client of engineering offices and management consultancies.

If you are working as a self-employed sustainability manager, then ask for at least a net daily rate of 900 € for a remote activity and 1100 € for on-site assignments, the whole thing of course plus expenses. There are no upper limits and in many countries daily rates of 1500 € and more are quite common. When making your fee demands, also consider that for every day on site you have to calculate at least one day of preparation and one day of follow-up. If you apply for such a management position as an employee, then you should ask for a fixed salary of 80,000 € per year plus a flexible performance component of 10,000 to 20,000 € per year.

Great, now you are a well-paid sustainability manager and use all your experience for the organization you are currently working for. You want to advance this organization and accompany it into a sustainable future and are enthusiastic and committed to the cause. You are now a fully trained sustainability manager full of pride and self-confidence and ready for the current situation, the new project, the new challenge. You have your backpack with all the tools and tools on your back (see appendix

12) that you have acquired in your previous professional life and with your further education as a sustainability manager.

Nevertheless, it may happen that despite everything, success still eludes you. If this is the case, then you simply do not have the right key to success in your hand.

Source Reference and Notes

1. Richard Bandler & John Grinder, Reframing—Neuro-linguistic programming and the transformation of meaning, Junfermann Verlag
2. e.g. at https://www.openproject.org/
3. Miyamoto Musashi, Gorin No Sho—The Book of Five Rings, RaBaKa Publishing

11

The Key to Your Success

I like to compare this situation with the following picture: You are standing in front of the building of an organization that you want to help to follow the path of sustainable development. You have all the skills and tools that we have discussed in this book in your backpack and are able to use them perfectly and successfully at any time. Fit and full of motivation, you now want to enter the building and start working. But when you try to unlock the door, you find that your key does not fit and the door remains locked.

This picture matches my own experience in sustainability projects quite well. I often wondered why, with all the factual knowledge on the topic of sustainability, there are not many more examples of successful sustainability management that I should have come across. Did I look too little and in the wrong places?

I don't think so, because anyone who has been dealing with the topic on a daily basis for more than fifteen years,

© The Author(s), under exclusive license to Springer-Verlag GmbH, DE, part of Springer Nature 2023
M. Wühle, *Making Sustainability Measurable*,
https://doi.org/10.1007/978-3-662-66715-6_11

as I have, gets a good overview. Of course, it is impossible to know all the successes in the field of sustainability, but I am in touch with important organizations and multipliers in the scene. Whether these are people that I meet in the context of my seminars and lectures or at conferences and events, I usually find seekers who often try in an admirable way to find the right way to a sustainable future for their company, their municipality or their association.

For example, the annual conference of the "Council for Sustainable Development in Germany" is an event that I attended regularly until a few years ago. In the lectures and symposia, methods and tools are also presented that each participant can apply as best practice for themselves. That is why many people with a connection to sustainability visit this event.

However, the vast majority of the participants of this and similar events belong, according to my perception, to the undecided and the seekers. In numerous conversations on the sidelines of many conferences of different organizations from the field of sustainability (networking is very important, as I said!) I have heard a wide range of questions. These range from: "I would like to …, but does that make sense?" to "We want this and that, but we don't know how to do it". This reminds me again of a movie where someone tries to fumble the key into the door lock in the dark, behind which the solution lies.

Again and again it is said in these places: "We are for sustainability goals, but the implementation does not work" and: "Where are the black sheep, where are the opponents of sustainable development?". Expressions like transformation, social transformation, transmitting worldwide, German leadership and many more seem to be the magic words that are used, but nobody understands (I don't either, by the way).

Then there are sentences like: "We have to implement economically what we want ecologically, preferably at the municipal level" and: "We have an institutional problem to coordinate the different concepts, e.g. grid expansion, goals, priorities, virtual power plants, decentralization", finally: "We have to think from the end, how we can achieve our sustainability goals".

All almost right and yet so far from the solution that it sounds helpless and quite clueless to me. I also searched for the right key and the right door for many years and did not find either or not at the same time. The longer I dealt with the topic of sustainability and its surprisingly great potentials for all personal and professional situations and questions, the more I realized that there must be a simple and easy solution here as well.

I started to talk to people from my network about their professional and private successes and failures, and tried to find the causes for both. At first I did not ask the right questions, but I did not give up and asked again and again and listened attentively to the answers. Do you remember the power of questions?

In doing so, the elements that decide success and failure became increasingly clear to me. Not too long ago, I had an inspiring thought while jogging through the forest, thinking about the successes and failures of previous projects of mine.

Back in my office, I formulated a thesis for the reasons of success and failure and wrote it down. In the meantime, I have often checked and confirmed this thesis based on my own experience and that of other people.

That is why the following is no longer a thesis for me, but the key to success:

THE KEY FOR SUCCESS OR FAILURE

1. If in the past projects and decisions, in the professional or private sphere, if developments have gone negatively or unsatisfactorily, then there was ALWAYS a break in the (for us NOT CONSCIOUS) structure of sustainability as the cause. One, two, or all three dimensions/pillars of sustainability were (UNCONSCIOUSLY) **not considered** or **unequally weighted.**

2. If, on the other hand, things went positively and successfully in the past, then there was ALWAYS a (UNCONSCIOUS) complete and harmonious structure of sustainability underlying it. In these cases, ALL THREE DIMENSIONS of sustainability were ideally balanced for the respective case.

> THE KEY TO SUCCESS IS THE CONSCIOUS, TARGETED AND SIMULTANEOUS USE OF ALL THREE DIMENSIONS OF SUSTAINABILITY AT THE RIGHT TIME.

This key to success has also been found by other people for themselves; sometimes it is also called the 'sweet spot'. When all three dimensions of sustainability overlap and complement each other in a balanced way, then the 'sweet spot' unfolds its optimal effect.

It was good for me that I came to this incredibly important insight by myself and only afterwards stumbled upon the sweet spot (see Fig. 11.1). Finding it was the well-known *blue-car-syndrome*. When you have chosen and bought a blue car, you suddenly see blue cars everywhere. It was similar for me with my key to success. I found more and more information, articles and concepts that were based on the same insight.

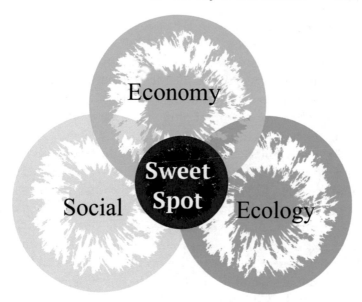

Fig. 11.1   Sweet spot

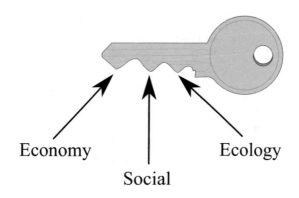

Fig. 11.2   The key to success

Now we know all the characteristics that make up the key to success (see Fig. 11.2). For it to fit the right lock, i.e. the right organization, it must have an individual shape for each organization. There is no universal key, but only individual shapes.

Our key to success in sustainability management always has three teeth, one for the economy, one for the social and societal responsibility and one for the ecology. However, the size and order of the teeth vary, they are different for each organization.

We sustainability managers have to make the key to success individually for each project and each organization. We can achieve this by analyzing the organization we want to help in detail. This way we find out the specific needs and bottlenecks and can file each of the three teeth so that it fits. For this part you have to take enough time, look very closely and file the key to fit. As with a real key, it does not help at all if it is 99% right. It will only be able to open the lock if it fits 100%.

That's why I like this picture so much. So take a close look at your organization, which you want to bring on the path of sustainability. Study all the data and documents you can get, talk extensively with the people involved, listen carefully, ask many constructive questions and get as accurate a picture of the organization as possible. If you talk to as many people in this organization as possible and ask the right questions, you will get the certainty and confidence to make a 100% fitting key for this organization. This information together will give you the picture of how your key for that particular organization must look exactly. You will then know how big you have to make the economic, the social/societal and the ecological tooth.

With this key you are almost sure of success and you can start your work with joy.

12

In the Right Place at the Right Time

We have almost reached the end of the book. We know the differences and similarities between the various forms of sustainability management and we share the insight that the key to success lies in the balanced application of all three dimensions of sustainability. We are convinced that ignoring or at least weakening even one of these dimensions will most likely cause our project to fail.

And yet this knowledge is not always enough to be sure that our respective project will succeed. May be it has also happened to you that your project, your measure, your undertaking failed or did not bring the success you had aimed for and that, although you—looking back—had taken into account all dimensions of sustainability in a balanced way, had the right key to success in your hand and had considered the respective peculiarities? This has also happened to me and I always wondered what the cause was.

© The Author(s), under exclusive license to Springer-Verlag GmbH, DE, part of Springer Nature 2023
M. Wühle, *Making Sustainability Measurable*,
https://doi.org/10.1007/978-3-662-66715-6_12

In my experience, you are in the right place, with the right key to success in your hand, but you are there at the wrong time. This has also happened to others. Do you remember Tolkien's "The Hobbit"? The key to the secret passage of the dwarves only worked on a certain day. And although the instructions read *"Stand by the grey stone when the thrush knocks and the last ray of sun on Durin's Day falls on the keyhole."*[1] the dwarves arrived much too early and had to wait for weeks for the right day. **It is therefore a time lock and our key to success only closes at the right time.**

You can recognize the wrong time by the fact that despite seemingly good or even ideal conditions, the willingness within the organization is not there to take the decisive step towards sustainability. If you listen attentively in the first conversations with the responsible people of the affected organization, you will quickly find out whether there are only isolated skeptics—they are always there, we know that—or whether the degree of doubt and indecision is too high to allow a real chance of success.

Do not try to convince the participants of the meaningfulness of your ideas in this case. We have discussed this in detail. You would not have any success with it. We do not even want to try to dissuade someone from their 'belief'!

In this case I advise you: **Stop before you have really started. You are at the right place at the wrong time.** Look for another organization that needs your help and for which it is the right time. After some time you can stop by the other organization again. Maybe the right time has come there too?

There is a right time for every project to implement. This is the case when all the necessary conditions are met and there is a broad majority among the participants who support the project. For sustainability projects, the right time is even more important than for 'normal' projects, because three dimensions have to be in the right relation to each other. Recognizing the right time is the supreme discipline in sustainability management. I am sure you will recognize the right time!

References and Notes

1. John Roland R. Tolkien, The Hobbit, Deutscher Taschenbuch Verlag dtv

13

Sustainability—The Future is Waiting for you

Let us take a brief look into the past before we devote ourselves entirely to the future. Right at the beginning of the book, we talked about the origins of the term sustainability and the need to weave sustainable structures firmly into our actions.

We understood that prosperity and a healthy environment are different aspects of the same thing. What Hans Carl von Carlowitz realized with his *wild tree cultivation* and its interaction with mining, with the population and his sovereign, was the fact that only sustainable economic activity is the basis for a generally positive development.

To transfer the original idea of sustainability into our time, we worked with several images. Similar to Steve Jobs, we combined different, seemingly incompatible approaches into a coherent overall picture. Following the famous image in which technology and art overlap in Steve Jobs' Apple universe and unite at the point where

© The Author(s), under exclusive license to Springer-Verlag
GmbH, DE, part of Springer Nature 2023
M. Wühle, *Making Sustainability Measurable*,
https://doi.org/10.1007/978-3-662-66715-6_13

great products are created, we introduced the image of a roundabout for sustainability.

In this roundabout of sustainability, the economic, the social/societal and the ecological dimension of the sustainability principle unite without collisions into a strong unit that finally takes the exit towards a sustainable development together.

We understood that these three dimensions do not hinder each other, but on the contrary, they even reinforce each other through positive interactions. We recognized that a result of $1 + 1 + 1 = 4$ comes out as a strong and powerful solution for many problems of our time.

Especially in the age of global climate change, holistic solutions are required. We got the initially vague notion that especially for the urgent and truly existential questions and problems that come along with the unstoppable global warming, the serious and professional application of the sustainability principle can be the ultimate solution.

Mind you, can be! It is like a magic spell that we learned as an apprentice Felix from our master Carlowitz, whose power we now know and therefore fear to apply it, to pronounce it, to work the magic.

But once we have uttered this magic spell, it will influence our entire life. Regardless of whether we have anything to do with the topic of sustainability in our professional life or not, this magic works and we cannot get rid of it.

We change our view of what is happening around us, we are very likely to change our own behavior, especially our consumption behavior, because with the gained knowledge we can easily calculate without a computer what we collectively cause with our standard behavior.

Everyone of us, really everyone, can contribute positively on his or her level to make our actions more sustainable. It sounds crazy, but maybe the global climate change and the accompanying negative effects such as droughts, floods and other extreme weather events are the missing strong impulse that drives the entire society towards sustainability and thus to a better society?

The principle of sustainability will completely change our life and our behavior patterns. In many countries, the realization is spreading that we are literally taking the air to breathe with the previous methods. Nevertheless, in addition to the industrialized countries, the developing and emerging countries also want to continue to grow economically in order to achieve the same standard of living that we have in Europe. The realization that this is only possible in the twenty-first century in harmony with the environment and the people affected is spreading increasingly. This is our great opportunity.

Sustainability and sustainability management is the happy combination of three dimensions that do not fit together in the 'normal' business world. You, as a well-prepared sustainability manager, are now able to connect and link the economic, the social/societal and the ecological components in a way that can produce amazing and sustainable results.

I wish you much success in your activities as a sustainability manager!

The requirements of this time will make you a sought-after project leader and manager. The financial success will be as well as the recognition by your fellow citizens and the project participants. The fact that you make a significant contribution to preserving our environment with your work, to help preserve the creation for our descendants, that will give you strength when the path should become rocky.

All your small and large successes in the field of sustainability will increasingly put you in a creative mode, in a flow that carries you and enables you to perform in ways that will seem incredible to you.

If you are sometimes discouraged because difficulties pile up in front of you, then do as I do and direct a sigh into the past to Master Carlowitz. That always helps me. He also had many difficulties to overcome and yet or precisely because of that he created a new system that met the needs of his time.

Sustainability. We can do that in our time too.

You will, just like me, meet wonderful people on this journey who, like you and me, are firmly convinced of the meaning of sustainability and who, like us, do their best to make our world more sustainable.

I have one more small request for you: Pass on what you have learned through this book. Pass on this knowledge and your own insights and treasures of knowledge that you will get over time.

Sustainability is a generational contract. Always remember that everything we do should be suitable for our grandchildren. Sustainable, that is.

I wish you much success and much fun. Take care. Servus.

Hohenlinden in July 2022, Michael Wühle.

Appendix

Disclaimer: The author assumes no liability for the timeliness, correctness, completeness or quality of the information, examples and tools provided. Liability claims against the author, which refer to damages of a material or immaterial nature caused by the use or non-use of the information presented or by the use of incorrect and incomplete information, are generally excluded.

Appendix 1: SWOT Analysis

With such an analysis, you can illustrate very nicely where your organization stands in terms of sustainability, what risks exist and where the fields of action are.

© The Editor(s) (if applicable) and The Author(s), under exclusive license to Springer-Verlag GmbH, DE, part of Springer Nature 2023
M. Wühle, *Making Sustainability Measurable*,
https://doi.org/10.1007/978-3-662-66715-6

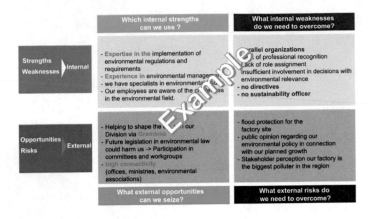

Which internal strengths can we use ?	What internal weaknesses do we need to overcome?
- Expertise in the implementation of environmental regulations and requirements - Experience in environmental management - we have specialists in environmental - Our employees are aware of the challenges in the environmental field.	- parallel organizations - Lack of professional recognition - Insufficient involvement in decisions with environmental relevance - no directives - no sustainability officer
- Helping to shape the law in our Division via Gremien - Future legislation in environmental law could harm us -> Participation in committees and workgroups - high connectivity (offices, ministries, environmental associations)	- flood protection for the factory site - public opinion regarding our environmental policy in connection with our planned growth - Stakeholder perception our factory is the biggest polluter in the region
What external opportunities can we seize?	What external risks do we need to overcome?

SWOT analysis

Appendix 2: The Temple of Sustainability

The sustainable orientation of an organization follows the structural design of a classic temple. The foundation of sustainability is the social responsibility that every organization and every company has to take on, and the knowledge of the ability of the system of sustainability to meet this responsibility. The three equally strong pillars of economy, social/cultural and ecology support the roof of sustainability: the sustainability management. Only the symmetry of this construction enables sustainable organizations and companies in real economic life.

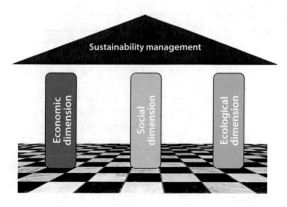

Temple of sustainability

Components of the System of Sustainability

1. Social responsibility, the foundation
2. Economic dimension of sustainability
3. Social dimension of sustainability
4. Ecological dimension of sustainability
5. Sustainability management

Appendix 3: Sustainability Criteria for Purchasing

The following list of criteria for sustainable purchasing is only exemplary and must be adapted specifically for each organization.

Criteria	Questions
Location	• Is the supplier local, nearby, or distant? • Is the supplier in a critical country in terms of human rights and labor practices? • Is the supplier in a critical country in terms of logistics, infrastructure, law, language, time zones?
Capacity	• Is the capacity of machines, premises, staff, and storage sufficient?
Reliability	• Has the supplier been reliable in the past in terms of delivery times and quality?
Technology/Skills	• Does the supplier have enough knowledge about his products? • Does the supplier have a system for quality assurance? • How old is the machinery?
Creditworthiness	• Is the financing of the production secured in the long term?
References	• Are there testimonials? • Are there ratings? • Are there certifications?
Synergy effects	• Are there contacts with the supplier through other companies?
Image/Reputation	• How high is the qualification of the staff? • Has the supplier implemented sustainability criteria? • Is there a sustainability officer?

Appendix 4: Fields of Action in (Municipal) Sustainability Management

The following lists serve to identify the respective fields of action of a municipality that are necessary for the establishment of a functioning sustainability management. The fields of action listed are exemplary.

Fields of action	Possible courses of action	Potentials
Sustainable procurement	Regional concept	A regional concept for the procurement of as many goods and services as possible from the region has a very positive impact on the value creation, jobs, environmental and climate protection
	Code of Conduct	Corruption prevention in municipal tenders increases the trust of the public
Inventory of energy consumption	Electricity, heat and cold per building	For organizations that are addressing this topic for the first time, savings of 30% and more are possible here with little effort and low costs

Fields of action	Possible courses of action	Potentials
	Load profiles	Enable the precise consumption analysis and are the basis for optimal energy efficiency and targeted use of renewable energies
Potential analysis for renewable energies and Biofuels	Energy-autonomous municipality	Energy autonomy by means of own and local sources is possible, if the question of energy storage (e.g. biogas) is considered and solved specifically for each municipality
Project development for the use of renewable Energies	Energy cooperative	The establishment of a citizen energy cooperative with municipal participation is a very good way to implement projects for the use of local energy sources quickly and easily. Funding from state/federal/EU can facilitate the financing and increase the return

Fields of action	Possible courses of action	Potentials
Sustainability criteria for investment measures	Tender management	Sustainability criteria are included into tenders for machines, plants and vehicles of the municipality. This can save considerable costs over the planned lifetime of the investment, achieve environmental and climate goals and guarantee compliance with social standards
Citizen models as an element of a sustainable municipal development	Citizen energy cooperatives	A cooperative as a grassroots democratic organization offers citizens of all age groups the opportunity to actively participate in a municipal energy transition and take responsibility for the municipality
Use of revenues from the use of Renewable energies	Village beautification	The municipal share of the revenues of a citizen energy cooperative can be used for municipal expansion measures, village beautification and support of regional associations and initiatives

Fields of action	Possible courses of action	Potentials
Sustainable Personnel marketing	Image building	Successful recruitment of skilled and managerial staff in competition with the economy by showing the strategic sustainability goals of the municipality (renewable energies, village beautification, infrastructure) and the associated attractive tasks and positions
Education	Municipal education planning	Increasing the attractiveness of the municipality through a school development and youth welfare planning with targeted involvement of the citizens
Generation-appropriate planning	Working group	Suggestions for achieving a high degree of accessibility in the municipality, dialogue between the different stakeholder groups

Fields of action	Possible courses of action	Potentials
Social networks	Partnerships	Establishment of partnerships with neighboring municipalities and/or abroad. Exchange of best practice on the topic of renewable energies and adaptation to climate change
Creation of CO_2-footprint and life cycle assessment	Energy self-sufficient municipality	Implementation of a zero-emission strategy with the aim of reducing the greenhouse gas emissions of the municipality by 95 percent by 2050 compared to 1990
Agreement of CO_2-reduction targets and -degradation paths	Citizen participation e.g. in the form of citizen energy cooperatives	Agree on goals for energy saving/energy efficiency/energy management (short-/medium-/long-term) Agreement on the share of locally generated renewable energy Agreement on the share of external green electricity providers (change when awarding new electricity supply, taking into account the notice periods)

Fields of action	Possible courses of action	Potentials
Analysis of local climate models and their possible impacts	Commission for the creation of a local climate model	Identification of risks for the municipality due to the consequences of global climate change Local climate models transfer the results of the global climate models to a small-scale level up to 50 km². Local peculiarities are taken into account
Risk analysis of the consequences of climate change	Adaptation measures	From the local climate model, a risk analysis (e.g. with SWOT) for the municipality can be derived. From this, forward-looking planning can be derived, e.g. for flood protection
Concept of a stable local ecosystem	Biodiversity	Also from the local climate model, target measures for forest conservation and biodiversity conservation can be derived. For municipalities, there are many possibilities of funding

Fields of action	Possible courses of action	Potentials
Resource-saving action	Road infrastructure and municipal buildings	Preferred use of recycled building material for new construction and renovation, while also examining the possible conversion to LED lights, Footpath and cycle path concept as an alternative to the car
Wastewater	Sewage treatment plant	Conversion/renovation/new construction of the municipal sewage treatment plant with a focus on self-generated electricity through photovoltaic system and biogas use in micro-CHP (self-generated electricity privilege)

Appendix 5: Benefits of a Sustainability Report

The benefits of reporting are manifold:

- **Customer attention:** Of course—the price is important. But it is not everything. In saturated markets and with comparable products, it is emotions that decide. Here, a responsible appearance can be a significant advantage over competitors. Because the more attentive the customers become, the more the differences to competitors count! More and more consumers base their purchasing decision on the sustainability of the producer or service provider. With a good sustainability report, we reach the desired attention of our customers as well as all other stakeholders.

- **Employee motivation:** Sustainability reports motivate the own workforce! They give courage, because they show that personal values can be reconciled with one's own economic activity. In addition, the positive feedback from colleagues, employees, friends and family brings new energy. The degree of identification with the own organisation usually increases noticeably, which in turn is helpful for the business activity of the organisation.

- **Better recruitment:** If your organisation credibly conveys through the sustainability report that it takes its social responsibility seriously and offers meaningful employment, you not only win the best minds, but also people who bind themselves sustainably (!) to the organisation.

- **Investor confidence:** Investors have always been interested in the opportunities and risks of a company. In addition to the economic, there are also environmental risks that the organization, the company has to cope with. Presenting these environmental risks in a sustainability report and not sweeping them under the carpet creates trust among potential investors. Often, a comprehensive sustainability report can achieve a better rating from investors and lenders. Sustainability reports also show the innovations and opportunities that lie in the vision of sustainable development. For listed companies, sustainability reports are the ticket for special sustainability/ethical funds, which have already developed well due to the increased attention of investors to sustainability and will do so even more in the future, in my opinion.

- **Improved access to political decision-makers:** Politicians also like to surround themselves with successful companies that embody future viability, responsibility and environmental protection. Especially in recent years, I notice this effect in an increasing degree. Apparently, many politicians have realized that in a globalized world, a new entrepreneurial approach must be found, which is given by the sustainability methodology. With your sustainability report, you can specifically draw the attention of politicians to yourself and thus gain useful contacts and access to political decision-makers.

- **Good understanding with authorities and neighborhood:** Sustainability reports provide a positive access to the respective organization through their open information. The stakeholders quickly notice that the sustainability reporting, especially according to GRI criteria, enables transparent and comparable information over the years on the core issues of the organization. This creates trust and is an important basis for an open dialogue and a general goodwill towards the organization. At the same time, it enables measures and projects of the organization with impacts on neighbors to be carried out faster.

- **Orientation in management:** Last but not least, sustainability reports provide clarity for yourself. The reporting is like a scan over your company, making successes and challenges obvious. A good sustainability report describes how a company wants to secure its future in the long term. The preparation of the reports therefore offers a good opportunity to examine the environment for ecological, social and economic opportunities and risks. In any case, the concise and comprehensive information of a sustainability report creates a completely new perspective on your own organization, which in turn offers new opportunities.

Appendix 6: Strategic Cornerstones for Sustainable Development in Municipalities

In the age of global climate change, the transition from fossil fuels to renewable energies, strategic cornerstones have to be defined for municipalities that enable sustainable development.

- **Involvement of citizens**   The promotion of participation and initiative of as many citizens as possible in all important issues of municipal politics with the aim that the people take their concerns in the community into their own hands. For this, the municipal administration must define clear responsibilities and accountabilities and communicate them within the municipality. The appointment of citizens to advisory boards that deal with issues related to municipal sustainability management is an effective way of promoting participation.
- **Resource consumption and energy production** Development of strategies and measures to reduce greenhouse gas emissions and conserve resources as much as possible. Promotion of the formation of municipal citizen energy cooperatives for the construction of energy generation plants from renewable and local sources.

- **Sustainability aspects in financial planning** L a r g e r investments are made exclusively taking into account the total costs during the lifetime. In the procurement sector, sustainability criteria relevant for awarding contracts are implemented, which will also lead to cost reductions. The use of innovative technologies that can reduce energy, fuel or the consumption of natural resources is specifically applied.

- **Sustainability guidelines** The goals and milestones of the municipal sustainability strategy are regularly communicated within the administration and to the citizens, associations, clubs and companies of the municipality. This creates a common understanding of the municipal vision of sustainability.

- **Role distribution** The involvement of as many citizens as possible is the declared goal of any municipal development. For this, however, the municipality has to take a pioneering role and create the appropriate framework conditions. Creating these prerequisites and structures and providing the relevant information is the task of the mayor, which cannot be delegated.

Appendix 7: Sustainable Balanced Scorecard

A balanced scorecard can illustrate a simple assessment of the degree of sustainability in an organization.

Finance			
Objective	Measured variables	Target values	Measures

Customers / Stakeholders			
Objective	Measured variables	Target values	Measures

Vision & Strategy

Environment / Climate			
Objective	Measured variables	Target values	Measures

Learning & Development			
Objective	Measured values	Target values	Measures

Processes			
Objective	Measured values	Target values	Measures

Sustainable Balanced Scorecard

The Sustainability Balanced Scorecard is based on a previously developed sustainability strategy and the underlying vision for the organization. For the five fields of finance, environment/climate, processes, learning & development and customers/stakeholders, goals are defined first. Metrics and target values specify and quantify the goals, for which measures are finally defined.

In order to take into account all three dimensions equally, I assign the color of the pillar of the temple of sustainability to the individual fields and measures, which is primarily or predominantly applicable for the measure. This allows to visually see whether our balanced scorecard is really balanced.

Appendix 8: Checklist—Mitigation of Climate Change

The following excerpt from a checklist according to the system *Sustainability. Now.*[*] gives a first overview of necessary adaptation measures. The complete checklist can be purchased at https://www.nachhaltigkeit-management.de/.

4	Ökologie	
4.3	Abschwächung des Klimawandels und Anpassung	
4.3.3	Anpassung an den Klimawandel	
	Hat die Organisation ein Risikomanagement aufgebaut? Berücksichtigung der zukünftigen globalen und örtlichen Klimaprognosen und Identifizierung der Risiken für die Organisation	○ Ja ○ Nein ☐ Nachweise sind beigelegt. Anlage:
	Berücksichtigt die Organisation die Auswirkungen des Klimawandels? Bei der Planung der Landnutzung, Flächennutzung und Gestaltung der Infrastruktur sowie der Instandhaltung	○ Ja ○ Nein ☐ Nachweise sind beigelegt. Anlage:
	Unterstützt die Organisation regionale Maßnahmen zur Reduzierung von Überflutungen? Dies beinhaltet den Ausbau von Feuchtgebieten zum Hochwasserschutz und Reduzierung der Flächenversiegelung in Stadtgebieten.	○ Ja ○ Nein ☐ Nachweise sind beigelegt. Anlage:
	Trägt die Organisation zur ökologischen Bewusstseinsschärfung bei? z.B. durch entsprechende Seminare und Fortbildungsmaßnahmen	○ Ja ○ Nein ☐ Nachweise sind beigelegt. Anlage:
	Werden Gegenmaßnahmen eingeleitet? Einleitung von Gegenmaßnahmen zu bestehenden oder zu erwartenden Auswirkungen. Beitrag im eigenen Einflussbereich, so dass Anspruchsgruppen Kompetenzen und Fähigkeit zur Anpassung aufbauen	○ Ja ○ Nein ☐ Nachweise sind beigelegt. Anlage:

Excerpt from the checklist on climate change

Appendix 9: GRI Standards

Once the reporting principles have been implemented, the contextual information and the material topics of the organizations have been identified, the sustainability report can be prepared in the structure of the GRI standards:

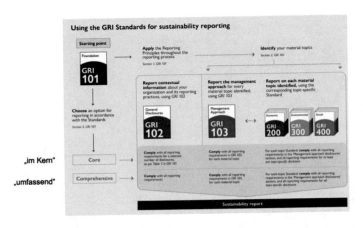

„im Kern"

„umfassend"

Schema GRI standards. (Source: www.globalreporting.org)

Elements of the GRI standards. (Source: www.globalreporting.org)

Appendix 10: Transformation from GRI –G4 to GRI Standards

Tool "Mapping G4 to the GRI Standards":
  https://www.globalreporting.org/standards/
resource-download-center

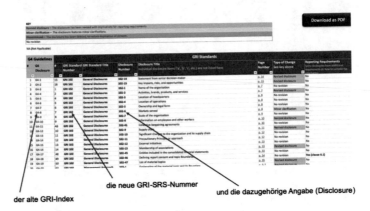

die neue GRI-SRS-Nummer

und die dazugehörige Angabe (Disclosure)

der alte GRI-Index

Transformation G4 to GRI standards. (Source: www.globalreporting.org)

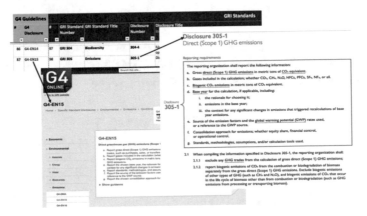

Example of transformation G4 to GRI standards. (Source: www.globalreporting.org)

Appendix 11: Links to Helpful Tools

The following links lead to tools and information in the field of sustainability. The tools are partly free of charge.

- VFU e. V. (Association for Environmental Management and Sustainability in Financial Institutions)—Calculation of environmental indicators, accounting of greenhouse gases https://vfu.de/2022/07/12/update-des-vfu-kennzahlenstandard-2022-auf-die-version-1-1/
- Key Performance Indicators for Environmental, Social & Governance Issues https://www.dvfa.de/fileadmin/downloads/Publikationen/Standards/KPIs_for_ESG_3_0_Final.pdf
- WeSustain—Software solutions for CSR management https://www.wesustain.com/
- Sustainability certification and non-financial reporting fororganizationsandcompanieshttps://www.nachhaltigkeit-management.de/zertifizierung-nachhaltigkeit-mess-bar-machen/
- GaBi—Life-cycle analysis and eco-balance https://gabi.sphera.com/international/index/
- PlusB Consulting—Sustainability management, energy consulting, funding consulting not only for SMEs https://nachhaltigkeit-management.de
- Environmental Pact Bavaria—Sustainability management for SMEs https://www.umweltpakt.bayern.de/umwelt_klimapakt/
- Environmental Pact Bavaria—Greenhouse gas balance/Carbon footprint https://www.umweltpakt.bayern.de/energie_klima/fachwissen/279/carbon-footprint
- Guide to the German Sustainability Code https://www.nachhaltigkeitsrat.de/wp-content/uploads/migration/documents/Leitfaden_zum_Deutschen_Nachhaltigkeitskodex.pdf

Appendix 12: The Backpack of the Sustainability Manager

The backpack of the sustainability manager contains first of all the tools and instruments of the basic version. Like any basic equipment, it should be supplemented and optimized by you with special tools over the years.

Tool backpack of the sustainability manager

- The Temple of Sustainability
- SWOT analysis
- Sustainability Balanced Scorecard
- Checklists for assessing sustainability
- Sustainability criteria for purchasing

- Arguments for a sustainability report
- Information of a sustainability report
- Our value system
- Sustainability management
- Sustainability strategy

The backpack of the sustainability manager is packed with care and system. A sustainability strategy and a related sustainability management is part of our daily work, which we master even in our sleep. Therefore, we pack these things first at the bottom and pull them out only when needed. On top of that our value system, which we always refer back to and adjust when necessary.

In the upper area are our checklists, with which we can check the degree of sustainability of an organization in all three pillars. There we also find the Sustainable Balanced Scorecard and the SWOT analysis, because we need these things almost always.

And at the very top, immediately accessible is the Temple of Sustainability. It is in this position, because we as sustainability managers have to remind ourselves again and again, how the system of sustainability is structured and what dependencies exist. Moreover, we need the temple always at the beginning, when we have the first conversations with a new customer/client/contractor.

Appendix 13: Recording List of Energy Consumers

With this simple list you can record the consumers in your household and get an immediate overview of where your energy guzzlers are hidden.

No.	Device	Type	Power per device in watts	Stand-by power per device in watts	Operating days per year	Operating hours per day	Stand-by hours per day	Energy consumption in kWh per year
1	PC	0815	300	5	365	2	22	24,09
							Total:	

Example: Your PC consumes according to manufacturer information 300 W in normal operation and 5 W in stand-by mode. On an average day, the PC is in operation for 2 hours, the rest of the time on stand-by.

The energy consumption is then calculated with: E = ((300 W x 2 h) + (5 W x 22 h)) x 365 d/1.000.000 = 24.09 kWh per year.

Source Reference and Notes

1. More information on ethical funds e.g. at: https://www.geld-welten.de/geldanlage/fonds/ethikfonds.html
2. https://www.globalreporting.org/standards/gri-standards-translations/gri-standards-german-translations-download-center/

Printed in the United States
by Baker & Taylor Publisher Services